Universitext

Editors

F.W. Gehring
P.R. Halmos
C.C. Moore

Robert B. Reisel

Elementary Theory of Metric Spaces

A Course in Constructing Mathematical Proofs

Springer-Verlag
New York   Heidelberg   Berlin

Robert B. Reisel
Loyola University of Chicago
Department of Mathematical Sciences
Chicago, Illinois 60626
U.S.A.

Editorial Board

F.W. Gehring
University of Michigan
Department of Mathematics
Ann Arbor, Michigan 48104
U.S.A.

P.R. Halmos
Indiana University
Department of Mathematics
Bloomington, Indiana 47401
U.S.A.

C.C. Moore
University of California at Berkeley
Department of Mathematics
Berkeley, California 94720

Library of Congress Cataloging in Publication Data

Reisel, Robert B.
  Elementary theory of metric spaces.

  (Universitext)
  Includes index.
  1. Functions of real variables. 2. Functional
analysis. 3. Metric spaces. I. Title.
  QA331.R45   514'.32   82-864
                                  AACR2

AMS Classifications: 26-01, 46-01

© 1982 by Springer-Verlag New York, Inc.
All rights reserved. No part of this book may be translated or reproduced in any form without written permission from Springer-Verlag, 175 Fifth Avenue, New York, New York 10010, U.S.A.

Printed in the United States of America

9 8 7 6 5 4 3 2 1

ISBN 0-387-90706-8 Springer-Verlag New York Heidelberg Berlin
ISBN 3-540-90706-8 Springer-Verlag Berlin Heidelberg New York

Preface

Science students have to spend much of their time learning how to do laboratory work, even if they intend to become theoretical, rather than experimental, scientists. It is important that they understand how experiments are performed and what the results mean. In science the validity of ideas is checked by experiments. If a new idea does not work in the laboratory, it must be discarded. If it does work, it is accepted, at least tentatively. In science, therefore, laboratory experiments are the touchstones for the acceptance or rejection of results.

Mathematics is different. This is not to say that experiments are not part of the subject. Numerical calculations and the examination of special and simplified cases are important in leading mathematicians to make conjectures, but the acceptance of a conjecture as a theorem only comes when a proof has been constructed. In other words, proofs are to mathematics as laboratory experiments are to science. Mathematics students must, therefore, learn to know what constitute valid proofs and how to construct them. How is this done? Like everything else, by doing. Mathematics students must try to prove results and then have their work criticized by experienced mathematicians. They must critically examine proofs, both correct and incorrect ones, and develop an appreciation of good style. They must, of course, start with easy proofs and build to more complicated ones. This is usually done in courses, like abstract algebra or real analysis, in the junior and senior years of college, but this is almost too late in the curriculum. Furthermore, with the increase in the number of computer applications studied in these and in other courses, it is becoming more difficult to find enough time for a critical study of proof techniques. This book is intended to provide a text for an earlier course that would emphasize proofs while teaching useful and important mathematics.

As I said, students learn to construct proofs by actually working out proofs. Therefore, this book is not a theoretical treatise of logic or proof theory but is an actual text on metric spaces. Instead of giving the proofs of the results, I ask the students to supply them, giving them hints where necessary. Naturally, I have given all the definitions and some indication of what the ideas mean so that the book is self-contained. The students should be "on their own" as much as possible, but an instructor should be available to help them over their difficulties and to offer constructive criticism of their efforts.

Originally, I had intended not to include any proofs in the book, but I have relented in a few cases where it seemed unreasonable to expect beginning students to be able to complete some complicated proofs. Also I have included an appendix in which I have given proofs of selected results. I have tried to include those proofs which illustrate unfamiliar techniques or involve concepts that might cause difficulty. Ideally, the students should not look at the proofs in the appendix until they have written their own proofs, or at least tried to do so. However, if there is not enough time to cover the essentials of the theory of metric spaces and also do all the proofs, the course could be accelerated by having the students read the proofs in the appendix and only try to construct proofs for the other results.

I think that the best way to use this book is in a seminar; I give some suggestions for this below. It could, however, be used in a lecture course where many of the proofs would be assigned to the students. It would be suitable as the text or as a supplementary text in courses in general topology, real analysis or advanced calculus.

Having emphasized that the goal of the book is to teach an understanding of proofs and their construction, I now say that this should be forgotten. Students should concentrate on the actual mathematical content and try to learn and understand it. As they work through the mathematics, they will be learning proof techniques in what seems to be an incidental manner. All that is necessary is that high standards be insisted on and incorrect or incomplete proofs not be accepted.

The only prerequisite for this book is that nebulous quality called "mathematical maturity". I take this to mean that the students should be at ease working with mathematical formulas and symbols and be serious about learning mathematics. Students who have completed calculus should have this maturity. The subject matter of the book is the elementary theory of metric spaces. I say "the elementary theory" because I have restricted the topics to those whose proofs can be constructed by students working primarily by themselves. Nevertheless, the book develops all the ideas of metric spaces that are needed in a course in real analysis of functions of a single variable. The book also gives the students experience with mathematics in an axiomatic setting.

The preliminary chapter - Chapter 0 - is a brief explanation of those points of logic that students must understand in order to construct proofs. At the end of this chapter I have given some carefully written proofs for a number of exercises of Chapter I so that the students can check their first attempts at constructing proofs. They should be able to go through this chapter by themselves. A course using the book would actually begin with Chapter I. The first three chapters cover the essential material that everyone should know in order to understand modern analysis, namely, the elements of set theory, particularly the concept of a mapping, the basic geometrical ideas of metric spaces, and the properties of continuous mappings. The remaining three chapters take up sequences and completeness, connectedness, and compactness. Although these last three chapters involve general metric spaces, they are slanted toward the space of real numbers, which is the most important elementary application of these ideas and the one that is most familiar to the students. I have not included any discussion of product spaces, which would allow these results to be extended to higher dimensional spaces, because I think that product spaces are studied more effectively in the context of a topology course. Moreover, inclusion of this material would involve more difficult proofs and would extend the book beyond what could be reasonably covered in a single course. Students who have completed this book should have no difficulty reading about this or other advanced topics in books on metric spaces or topology. Mathematical induction is used in a few places, so I have included an appendix that briefly explains it.

For the past fifteen years I have used versions of this book in a seminar course. The advantage of a seminar for this course is that students not only construct proofs but also have an opportunity to examine critically the proofs presented by other students, some of which are probably incorrect. I usually have about ten students in the seminar and the course meets two hours a week for a semester. Assignments are made so that each student's report takes about a half hour. The students are told to write out their work completely, not using any other books. I am available if they run into difficulties, but I limit my aid to suggestions as to how to proceed. At the start of the course I ask the students to show me their work before presenting it in the seminar so that I can catch any gross errors or make stylistic suggestions. Later in the course the students should have gained enough confidence and experience so that it is no longer necessary for them to show me their work. The students who are not making reports take notes as in any course, and if there is material in the presentations that they do not understand, they ask about it. It is important that the students be encouraged to raise objections and questions because this is the way that their critical sense is developed. I am careful not to raise objections myself unless no one in the class has caught the error. If there are mistakes, attempts should be made to correct them in the class with the other students making suggestions. If they cannot be corrected in a reasonable period of time, the work can be postponed to the next meeting. It might not be possible to cover the entire book, so I have made some suggestions in the various chapters as to what might be omitted.

I want to mention a few technical points about the book. Results are numbered consecutively in each section with a pair of numbers, the first denoting the section and the second the particular result; thus, Theorem 3.2 is the second result of Section 3. Within a chapter a reference to a result is made by citing this number. A reference to a result in another chapter is made by including the chapter number; thus, Theorem II 3.2 is the second result of Section 3 of Chapter II. Definitions are not set off in separate statements, but the word being defined is underlined. It is a pecularity of mathematical style that definitions are expressed by using the word "if" rather than the phrase "if and only if," although the latter is actually meant. I have followed this convention. Those results whose proofs can be found in the appendix are marked with an asterisk (*). Finally, the symbol "[]" is used to indicate the end of a proof.

As I mentioned above, this book has reached its present form only after a considerable period of time and experimentation. I would like to list the names of all my colleagues who have used forms of this book over the years, but I am sure that I would forget to include some of those who left Loyola University years ago. Therefore, I will just express my thanks to all my colleagues, both past and present, for their suggestions, support and encouragement. I would also like to thank the many students who suffered through the seminar as it developed into its present form. Their successes and failures helped me to learn what can be reasonably expected in a course of this kind. Lastly, I want to thank my wife for her patience, encouragement and love. She was and is indispensable.

CONTENTS

Preface.	v
Chapter 0. Some Ideas of Logic.	1
Chapter I. Sets and Mappings.	14

    1. Some Concepts of Set Theory. 2. Some Further Operations on Sets. 3. Mappings. 4. Surjective and Injective Mappings. 5. Bijective Mappings and Inverses.

Chapter II. Metric Spaces.	34

    1. Definition of Metric Space and Some Examples. 2. Closed and Open Balls; Spheres. 3. Open Sets. 4. Closed Sets. 5. Closure of a Set. 6. Diameter of a Set; Bounded Sets. 7. Subspaces of a Metric Space. 8. Interior of a Set. 9. Boundary of a Set. 10. Dense Sets. 11. Afterword.

Chapter III. Mappings of Metric Spaces.	52

    1. Continuous Mappings. 2. Continuous Mappings and Subspaces. 3. Uniform Continuity.

Chapter IV. Sequences in Metric Spaces.	59

    1. Sequences. 2. Sequences in Metric Spaces. 3. Cluster Points of a Sequence. 4. Cauchy Sequences. 5. Complete Metric Spaces.

Chapter V. Connectedness.	69

    1. Connected Spaces and Sets. 2. Connected Sets in R. 3. Mappings of Connected Spaces and Sets.

Chapter VI. Compactness	75

    1. Compact Spaces and Sets. 2. Mappings of Compact Spaces. 3. Sequential Compactness. 4. Compact Subsets of R.

Afterword.	85
Appendix M. Mathematical Induction.	89
Appendix S. Solutions.	91
Index.	119

Chapter 0:
Some Ideas of Logic

The dependence of mathematics on logic is obvious. We use reasoning processes in mathematics to prove results, and logic is concerned with reasoning. However, when someone studies mathematics, he does not first study logic in order to learn to think correctly. Rather, he jumps into mathematics, perhaps with high school geometry, and learns to prove things by actually doing proofs. Logic comes into his education only when there seems to be something doubtful or obscure that needs clarification. A person who is working with some parts of mathematics, like foundational studies, where common sense does not provide enough precision, needs the more finely-tuned results of logic. However, in most of the undergraduate mathematics courses you can get by with informal reasoning and common sense. There are a few exceptions. It is necessary to have a clear understanding of some of the vocabulary of logic as used in mathematics and of the logic underlying the idea of a mathematical proof. I will try to clarify some of these points in this chapter. You should read it over quickly and refer back to it when the need arises. At the end of the chapter I will give you an opportunity to use what you have learned by asking you to construct some simple proofs.

I want to start by examining the meanings of the logical connectives. These are the words or symbols that are used to modify or combine sentences. By a sentence I mean a declarative sentence, one for which it makes sense to say that it is true or false. In this sense a question is not a sentence. To avoid misunderstandings, I will use the word "statement" instead of "sentence." Therefore, in what follows a statement is a declarative sentence, one which is either true or false. In logic the meaning of the statement is not important, only its "truth value" counts, that is, we speak only of its truth or falsity.

One way to modify a statement is to negate it. If "P" is a statement, the negation of "P", written "not-P", is the statement that is true when "P" is false and false when "P" is true. This can be expressed concisely by a truth table:

P	not-P
T	F
F	T

where "T" stands for "true" and "F" for "false". In such a table each line shows the connection between the truth values of "P" and "not-P". For example, the first line asserts that when "P" is true, "not-P" is false. Of course, if "P" is a statement that is written out, "not-P" is expressed by the appropriate grammatical construction. For example, if "P" is the statement, "It is raining," then "not-P" is the statement, "It is not raining."

There are a number of ways that two statements can be combined to give a new statement. Since we are only concerned with the truth or falsity of statements, we merely have to list the conditions under which the new statement will be true or false, depending on the truth or falisty of the original statements. In particular, the connectives "and" and "or" are defined by the following truth tables:

P	Q	P and Q	P or Q
T	T	T	T
T	F	F	T
F	T	F	T
F	F	F	F

Note that the statement "P or Q" is defined to be true if one or the other or both of the statements "P" or "Q" are true. This is not always the sense in which "or" is used in everyday speech. Sometimes "or" is used like this, but at other times "or" is used in the sense that "P or Q" is true when precisely one of the "P" or "Q" is true, but not when they are both true. It is to avoid such ambiguity that we must agree on the use of the word "or". In mathematics "or" is used in the way given by the truth table.

Most mathematical theorems have the logical form "If ..., then ... ." For example, "If f is a differentiable function, then f is continuous." It is useful to have a symbol for such a statement. I will write "$P \to Q$" for "If P, then Q". This is used particularly when "P" and "Q" are mathematical formulas. To give the precise meaning to such a statement I must list its truth table.

P	Q	$P \to Q$
T	T	T
T	F	F
F	T	T
F	F	T

The last two lines are a little surprising or mysterious. In ordinary speech we do not say "If P, then Q" when we know that "P" is false, so there is really no conflict between this strict logical usage and everyday speech. Note that the statement about functions given above is a true statement. If I gave you a function which is not differentiable but is continuous, this would not invalidate the statement. This is an illustration of the third line of the truth table.

Experience has shown that the above truth table is the most useful way of defining a statement of the form "If P, then Q".

An important rule of logic follows from this truth table. Suppose you know that "P → Q" is true and that "P" is true. This can only happen in the first line of the truth table, so you can conclude that "Q" is also true. This rule is called "detachment", that is, if "P → Q" and "P" are both true, then "Q" is also true.

A final logical connective is "P if and only if Q", which is often shortened to "P iff Q". When it is useful, I will use the symbol "P ↔ Q" for "P iff Q". The truth table is

P	Q	P ↔ Q
T	T	T
T	F	F
F	T	F
F	F	T

Note that "P ↔ Q" is true when "P" and "Q" have the same truth value and false when they have different truth values. It is easy to see that "P ↔ Q" means the same as "(P → Q) and (Q → P)". To show this I will write the truth table for this last statement:

P	Q	P → Q	Q → P	(P → Q) and (Q → P)
T	T	T	T	T
T	F	F	T	F
F	T	T	F	F
F	F	T	T	T

The last column is obtained by applying the definition of "and" to the third and fourth columns. Now observe that the truth values for the last column are exactly the same as those of "P ↔ Q". Since we are only concerned with the truth values of a statement, we can say that "(P → Q) and (Q → P)" and "P ↔ Q" are logically equivalent. (Two statements will be called logically equivalent if their truth tables are identical.) Usually, when you have a statement of the form "P ↔ Q" to prove, you will find it necessary to use this result to split the work into two parts, namely, "P → Q" and "Q → P", and then prove each part separately.

As I mentioned, most mathematical theorems have the logical form "P → Q". To prove such a statement, you only have to show that when "P" is true, then "Q" must be true. This will show that the only case in which "P → Q" is false (the second line of the truth table) cannot occur. Therefore, in the proof you assume that "P" is true and then try to deduce that "Q" is true. Your proof will have the following form:

   Given: P
   Prove: Q
   Proof:

The "Proof" will be a sequence of steps that lead from "P" (and previously known results) to "Q".

Sometimes a proof is easier to construct by proceeding indirectly. One such process uses the "contrapositive" of the statement you want to prove. The contrapositive of "P → Q" is the statement "not-Q → not-P". (This should not be confused with the converse of "P → Q", which is "Q → P".) The following truth table shows that the contrapositive of "P → Q" is logically equivalent to "P → Q".

P	Q	not-Q	not-P	not-Q → not-P
T	T	F	F	T
T	F	T	F	F
F	T	F	T	T
F	F	T	T	T

(Compare the last column with the truth table for "P → Q".) Therefore, if you have to prove a result of the form "P → Q" and you want to do so indirectly, you would write:

Given: not-Q

Prove: not-P

Then you will try to find a sequence of steps that lead from "not-Q" (and previously known results) to "not-P".

A second kind of indirect proof is "proof by contradiction". Here to prove "P → Q" you assume that this statement is false and try to derive a contradiction. To say that "P → Q" is false means that "P" is true and "Q" is false. (See the second line of the truth table for "P → Q".) Your proof will have the form:

Given: P

       not-Q

Prove: a contradiction

Then you will try to find a sequence of steps that lead from "P" and "not-Q" to a contradiction. Usually you do not know ahead of time what the contradiction will be, so you are working a little blindly. However, you are starting with two statements in the "given", so you have more to work with. The logical basis of "proof by contradiction" is the following. Let "R" be the statement

("P → Q" is false) → contradiction

You have proved that "R" is true. Since a contradiction is a false statement, this means that you must have the fourth line of the truth table for the statement "R". This line gives that the statement ("P → Q" is false) is actually false, so "P → Q" is true, as you wanted to prove.

One technical point should be observed here. In the indirect proofs you have to write the negations of various statements. Many times you will have a universal statement, that is, one that says that all the objects of a class have a certain property, for example, "All differentiable functions are continuous." The negation of such a statement says that there is an object of the class that does not have the property. The negation of the statement in the example is, "There is a differentiable function that is not continuous." Similarly, the negation of a particular statement, that is, one that says that some object of the class has a property, is a universal statement, saying that no object of the class has that property. Sometimes it is difficult to decide exactly how to write such negations, but usually common sense will suffice. In the more confusing cases you need further techniques of logic, which I will not consider here.

Before continuing with this chapter you should now go to Section 1 of Chapter I and try to work out in detail the proofs of the exercises. After you have done so, compare your proofs with those that follow below. It is possible that your proofs will not agree completely with mine. If this is the case, try to decide if yours are correct anyway. You should show your work to your instructor so that he can correct any errors you might have made and comment on your style.

WORKED EXERCISES OF SECTION 1, CHAPTER I

1. Prove: $A \subseteq B$ iff $A \cup B = B$.

The proof consists of two parts because of the "iff". First prove "$A \subseteq B \to A \cup B = B$". To prove the equality of two sets you must prove that an element of the first set is in the second and vice versa, so this part of the proof will itself consist of two parts. In the first of these two parts you have to prove "$x \in A \cup B \to x \in B$" and in the second, "$x \in B \to x \in A \cup B$".

(i) Given: $A \subseteq B$

$x \in A \cup B$

Prove: $x \in B$

Proof: 1. $x \in A \cup B$ (Given)

2. $x \in A$ or $x \in B$ (Definition of union)

3. $A \subseteq B$ (Given)

4. $x \in A \to x \in B$ (Definition of $\subseteq$)

5. $x \in B$ (Steps 2 and 4)

Notice that in step 2 if $x \in B$, you are finished, while if $x \in A$, Step 4 and the rule of detachment give $x \in B$.

(ii) Given: $A \subseteq B$

$x \in B$

Prove: $x \in A \cup B$

Proof: 1. $x \in B$ (Given)

2. $x \in A \cup B$ (Definition of union)

Next prove "$A \cup B = B \to A \subseteq B$". To prove "$A \subseteq B$", you must prove "$x \in A \to x \in B$".

Given: $A \cup B = B$

$x \in A$

Prove: $x \in B$

Proof: 1. $x \in A$ (Given)

2. $x \in A \cup B$ (Definition of union)

3. $A \cup B = B$ (Given)

4. $x \in B$ []   (Definition of equality)

(The symbol "[]" indicates the end of the proof.)

<u>Comment</u>. Note how I have written the "Given" and the "Prove" so as to be as explicit and as definite as possible. A common error is to merely repeat the statement of the problem in the "Given" and the "Prove". For example, in the last part of the above proof, you might write:

Given: $A \cup B = B$

Prove: $A \subseteq B$

If you do this, then it is necessary to decide how you prove that $A \subseteq B$. Of course, you would prove this as I did, but your "Given" and "Prove" were not as explicit as they could be.

2. Prove: $A \cap (B \cup C) = (A \cap B) \cup (A \cap C)$

The proof will consist of two parts because you have to prove the equality of sets.

(i) Given: $x \in A \cap (B \cup C)$

Prove: $x \in (A \cap B) \cup (A \cap C)$

Proof:
1. $x \in A \cap (B \cup C)$ (Given)
2. $x \in A$ and $x \in B \cup C$ (Definition of $\cap$)
3. $x \in A$ and $(x \in B$ or $x \in C)$ (Definition of $\cup$)
4. $(x \in A$ and $x \in B)$ or $(x \in A$ and $x \in C)$

(Here you should think about the meaning of Step 3. It says that x is in A and at the same time x is in one of B or C. You should think about this long enough to see that this is also what Step 4 says. This can also be done more formally by noting that Step 3 has the logical form "P and (Q or R)" and Step 4 has the logical form "(P and Q) or (P and R)". You could construct truth tables for each of these to see that they are logically equivalent.)

5. $(x \in A \cap B)$ or $(x \in A \cap C)$ (Definition of $\cap$)
6. $x \in (A \cap B) \cup (A \cap C)$ (Definition of $\cup$)

(ii) Given: $x \in (A \cap B) \cup (A \cap C)$

Prove: $x \in A \cap (B \cup C)$

Proof:

The proof of this part just consists of the steps of the proof of part (i) in reverse order. You should write it out yourself. []

3. Prove: $A \subseteq B$ iff $cA \supseteq cB$

(i) Given: $A \subseteq B$

$x \in cB$

Prove: $x \in cA$

Proof:
1. $x \in cB$ (Given)
2. $x \notin B$ (Definition of complement)
3. $A \subseteq B$ (Given)
4. $x \in A \rightarrow x \in B$ (Definition of $\subseteq$)
5. $x \notin B \rightarrow x \notin A$ (Contrapositive of 4)
6. $x \notin A$ (Detachment, Steps 2, 5)
7. $x \in cA$ (Definition of complement)

Note that Step 5 uses the contrapositive of Step 4. This is not an indirect proof, but it uses the logical equivalence of a statement and its contrapositive.

(ii) Given: $cA \supseteq cB$
$x \in A$

Prove: $x \in B$

Proof:
1. $x \in A$ (Given)
2. $x \notin cA$ (Definition of complement)
3. $cA \supseteq cB$ (Given)
4. $x \in cB \rightarrow x \in cA$ (Definition of $\supseteq$)
5. $x \notin cA \rightarrow x \notin cB$ (Contrapositive of 4)
6. $x \notin cB$ (Detachment, Steps 2, 5)
7. $x \in B$ [] (Definition of complement)

Comment. You might notice that the second part of the proof is similar to the first. This might suggest to you that you could avoid writing the second part of the proof by referring to the first part in a certain way. This idea is correct and I will sketch out how it could be done. It is based on the fact that for any set A, $c(cA) = A$. (You should try to prove this.) In the first part you proved that if A and B are any sets and if $A \subseteq B$, then $cA \supseteq cB$. Now just write out this statement, replacing A by cB and B by cA. When you use the result about the double complement, you will get the statement that is proved in the second part of the above proof.

4. Prove: $c(A \cup B) = cA \cap cB$ and $c(A \cap B) = cA \cup cB$

As usual you have to prove that every element in one of the sets is in the other and vice versa. The proofs will depend on the way that logic tells us how to negate "or" and "and" statements. The negation of the statement "P or Q" is the statement "(not-P) and (not-Q)" and the negation of "P and Q" is "(not-P) or (not-Q)." If you are not familiar with these rules, you can easily prove them by writing the appropriate truth tables. Here is the proof of "$c(A \cup B) = cA \cap cB$; as usual it consists of two parts.

(i) Given: $x \in c(A \cup B)$

    Prove: $x \in cA \cap cB$

    Proof: 1. $x \in c(A \cup B)$ (Given)

              2. $x \notin A \cup B$ (Definition of complement)

              3. not-($x \in A$ or $x \in B$)   (Definition of union)

              4. $x \notin A$ and $x \notin B$ (See above)

              5. $x \in cA$ and $x \in cB$ (Definition of complement)

              6. $x \in cA \cap cB$ (Def. of intersection)

(ii) Given: $x \in cA \cap cB$

    Prove: $x \in c(A \cup B)$

The proof here just reverses all the steps of the preceding proof. You should write it out yourself. This proves the first of the two equations of this exercise.

I am not going to write out the proof of the second equation, $c(A \cap B) = cA \cup cB$, because it follows the same pattern as the proof of the first equation. Write out the proof yourself. Another way of proving this equation is by using the trick I discussed in the comment to Exercise 3. Rewrite the first equation by replacing "A" by "cA" and "B" by "cB" and then manipulate the result to get the second equation. []

5. Prove that the empty set is a subset of every set.

Take an arbitrary set A. As with any problem of set inclusion, you start with an element of the supposed subset and try to prove that it is in the larger set, that is, if x ε ∅, then x ε A. I will do this by an indirect proof, using the contrapositive, if x ∉ A, then x ∉ ∅.

    Given: x ∉ A
    Prove: x ∉ ∅
    Proof: 1. x ∉ ∅   []   (Definition of ∅)

Note that the proof is extremely short because of the use of the contrapositive. Sometimes a careful analysis of a problem can reduce it to a triviality.

6. Prove that there is only one empty set.

This will be a proof by contradiction. Assume that the result is false, that is, that there is more than one empty set, and derive a contradiction.

    Given: ∅ is an empty set
             ∅' is an empty set
             ∅ ≠ ∅'
    Prove: a contradiction
    Proof: 1. ∅ ⊆ ∅'   (Exercise 5 with ∅ as the empty set and ∅' as the set A)
           2. ∅' ⊆ ∅   (Exercise 5 with ∅' as the empty set and ∅ as the set A)
           3. ∅ = ∅'   (Definition of equality)
           4. ∅ ≠ ∅'   (Given)
           5. contradiction   (Steps 3 and 4)   []

In this proof the contradiction involves an assumption of the "Given", but in other proofs the contradiction might involve some previously proved theorem.

This concludes Chapter 0. You have learned some of the basic ideas of logic and proof techniques. Now the course begins in earnest. Go to Section 2 of Chapter I and begin to work. Good luck!

Chapter I:
Sets and Mappings

In this chapter you will learn some of the basic ideas of set theory, particularly those associated with mappings or functions. These will be used extensively in the following chapters and, indeed, in all of mathematics. This is not a systematic treatment of set theory. It omits many essential ideas, such as, infinity, cardinal and ordinal numbers, and foundational problems, that are not directly needed in this book. I assume that you have had some experience with the use of sets, so I do not give many examples of the familiar terms. Of course, you should not look in other books to find proofs of the theorems and the exercises of this chapter, because that would defeat the purpose of the book. However, if you want to see further discussion of the topics taken up here, you can look at books in set theory.

1. <u>SOME CONCEPTS OF SET THEORY</u>.

In the following it is assumed that there is given some universal set U and that all the sets that will be considered are composed of elements of U. If A is a set, the statement "a is an element of A" is denoted by "a ε A". Two sets A and B are defined to be equal if they have the same elements, that is, for every element x, x is an element of A iff x is an element of B. This can be written in symbols

as follows:
$$(\forall x)(x \in A \leftrightarrow x \in B).$$
("$\forall$" is the logical symbol for "for every" and "$\exists$" for "there exists". These expressions are called "quantifiers".) A set A is a __subset__ of a set B, written "$A \subseteq B$" or "$B \supseteq A$", if the following statement is true:
$$(\forall x)(x \in A \rightarrow x \in B),$$
that is, every element of A is an element of B. It follows that $A = B$ iff $A \subseteq B$ and $B \subseteq A$.

Let $p(x)$ be an expression which becomes a statement when any element of U is substituted for x. The symbol "$\{x \mid p(x)\}$" will denote the set of all the elements of U for which $p(x)$ is true when the element is substituted for x. If A is a set, "$\{x \mid x \in A$ and $p(x)\}$" or "$\{x \in A \mid p(x)\}$" will denote the set of all the elements of A for which $p(x)$ is true when the element is substituted for x. The concepts of __union__, __intersection__, __difference__, and __complement__ of sets are defined, respectively, as follows:
$$A \cup B = \{x \mid x \in A \text{ or } x \in B\},$$
$$A \cap B = \{x \mid x \in A \text{ and } x \in B\},$$
$$A - B = \{x \mid x \in A \text{ and } x \notin B\},$$
$$cA = U - A = \{x \mid x \notin A\},$$
where "$x \notin A$" means "not-$(x \in A)$". (In case it is necessary to specify the universe relative to which complements are taken, the symbol "$c_U(A)$" will be used instead of "cA".)

The empty set, which is a set with no elements, is denoted by "$\emptyset$". You will prove in Exercise 6 of this section that there is only one empty set and this will justify the use of the definite article "the" when referring to the empty set.

EXERCISES.

In the following problems you are asked to prove only a few of the many properties of sets and their operations. This should give you enough experience so that you will be able to prove any other properties that might come up later.

1. Prove: $A \subseteq B$ iff $A \cup B = B$.

(Hint. The proof consists of two parts. First assume $A \subseteq B$ and prove that $A \cup B = B$. To prove this equality you must prove that every element of $A \cup B$ is in B and that every element of B is in $A \cup B$. Secondly, assume that $A \cup B = B$ and prove $A \subseteq B$.)

2. Prove: $A \cap (B \cup C) = (A \cap B) \cup (A \cap C)$.

(Hint. Your proof might get quite involved with statements that contain several "and's" and "or's". Try to be as precise as your experience with logic will allow.)

3. Prove: $A \subseteq B$ iff $cA \supseteq cB$.

4. Prove: $c(A \cup B) = (cA) \cap (cB)$ and $c(A \cap B) = (cA) \cup (cB)$.

5. Prove that the empty set is a subset of every set.

(Hint. As with any proof of inclusion, you have an "If ..., then ..." statement. Try an indirect proof.)

6. Prove that there is only one empty set.

(Hint. Assume that there are two empty sets and use Exercise 5.)

2. SOME FURTHER OPERATIONS ON SETS.

Given two sets A and B, which might be identical, their Cartesian product, written "AxB", is defined to be the set of all ordered pairs (a,b), where a ε A and b ε B. (Two ordered pairs (a,b) and (c,d) are equal iff a = c and b = d.) Note that if R is the set of real numbers, then RxR is just the set of points in the plane of analytic geometry. (I will always use the letter "R" as the symbol for the set of real numbers. In Chapter II it will actually be used in a slightly more special sense, but this will cause no confusion.) Cartesian products of more than two, but still finitely many, sets are defined in an obvious way by using ordered triples, quadruples, etc.

A _family_ of sets indexed by a nonempty set Γ is a system consisting of a set A_γ for each $\gamma \in \Gamma$. (Γ and γ are, respectively, the upper case and lower case forms of the Greek letter "gamma". There is no deep reason why I am using Greek letters for the indexing set and for the index, but it helps in avoiding confusion with the sets of the family.) The family will be denoted by the symbol "$(A_\gamma \mid \gamma \in \Gamma)$". Note that I am using parentheses, not braces, in the symbol for a family. If $\Gamma = \{1, 2, \ldots, n\}$ or if $\Gamma = \underline{N}$, the set of positive integers, then a family indexed by Γ is just a finite or infinite sequence of sets. It is important to notice that the sets in a family need not be distinct, that is, it is possible that for two different elements α and β in Γ, the sets A_α and A_β might be equal. (See the comment below.)

If $(A_\gamma \mid \gamma \in \Gamma)$ is a family of sets, the _union_ and the _intersection_ of the family are defined by

$$\bigcup_{\gamma \in \Gamma} A_\gamma = \{x \mid \text{there is at least one } \gamma \text{ in } \Gamma \text{ with } x \in A_\gamma\}$$

$$\bigcap_{\gamma \in \Gamma} A_\gamma = \{x \mid x \in A_\gamma \text{ for every } \gamma \text{ in } \Gamma\}.$$

If $\Gamma = \{1, 2, \ldots, n\}$, the union is written "$A_1 \cup A_2 \cup \ldots \cup A_n$" and similarly for the intersection. Note that the above definitions can be expressed as follows:

$$x \in \bigcup_{\gamma \in \Gamma} A_\gamma \quad \text{iff} \quad (\exists \gamma)(x \in A_\gamma)$$

$$x \in \bigcap_{\gamma \in \Gamma} A_\gamma \quad \text{iff} \quad (\forall \gamma)(x \in A_\gamma).$$

<u>Comment</u>. There is an important distinction between a family of sets and a set (or collection) of sets. For example, let $\Gamma = \{1, 2, 3\}$ and A_1, A_2, and A_3 be sets with $A_2 = A_3$. Then the family $(A_\gamma \mid \gamma \in \Gamma)$, which can also be written as (A_1, A_2, A_3), is a sequence of three sets, two of which happen to be equal. We can speak of the first or second or third set in the sequence. On the other hand, $\{A_\gamma \mid \gamma \in \Gamma\}$ is the set $\{A_1, A_2, A_3\}$, whose elements are themselves sets. By the definition of equality of sets, this is equal to $\{A_1, A_2\}$ or to $\{A_2, A_1\}$. So this is a set of only two elements, and, moreover, it makes no sense to speak of the first or second element. It is always possible to consider a set of sets as a family of sets by choosing an indexing set and assigning an index to each set. If you were given, for example, the set of sets $\{A,B\}$, you could take Γ to be $\{1,2\}$, or any two-element set, and let $A_1 = A$ and $A_2 = B$. Note that the union of this set of sets and the union of the family are the same, that is, $A \cup B = \bigcup_{\gamma \in \Gamma} A_\gamma$. A similar remark holds for intersections.

<u>EXERCISES</u>. Proofs of exercises and theorems marked with an asterisk (*) can be found in Appendix S.

1.* Prove: $(A \times B) \cap (C \times D) = (A \cap C) \times (B \cap D)$.

2.   Prove: $(A \times B) \cup (C \times D) \subseteq (A \cup C) \times (B \cup D)$. (See the comment at the end of this section.)

(Hint. First prove that if P, Q, R and S are statements, then
[(P and Q) or (R and S)] → [(P or R) and (Q or S)]. To do this, assume that [(P and Q) or (R and S)] is true and prove that the other expression must be true. Remember that an "or" statement is true if

either constituent is true and that an "and" statement is true if both constituents are true.)

3. Prove: If $A \subseteq A'$ and $B \subseteq B'$, then $A \times B \subseteq A' \times B'$.

4.* Prove: $c(\bigcup_{\gamma \in \Gamma} A_\gamma) = \bigcap_{\gamma \in \Gamma} (cA_\gamma)$ and $c(\bigcap_{\gamma \in \Gamma} A_\gamma) = \bigcup_{\gamma \in \Gamma} (cA_\gamma)$.

5. Prove: $B \cap (\bigcup_{\gamma \in \Gamma} A_\gamma) = \bigcup_{\gamma \in \Gamma} (B \cap A_\gamma)$.

Comment. Many times it is necessary to show that a certain general result is not always true. This is done by producing an example in which the general result is false. In Exercise 2 you could ask, for example, if the inclusion could be replaced by an equality. To show that it cannot, you could produce specific sets A, B, C and D for which $(A \times B) \cup (C \times D)$ does not equal $(A \cup C) \times (B \cup D)$. Such an example of sets is called a <u>counterexample</u>. In this exercise you could take $A = \{1\}$, $B = \{2\}$, $C = \{3\}$, and $D = \{4\}$; each set is a singleton, that is, a set with only one element, and all the sets are different. Then $A \times B = \{(1,2)\}$ and $C \times D = \{(3,4)\}$, both of which are singletons, and $(A \times B) \cup (C \times D) = \{(1,2), (3,4)\}$. On the other hand, $A \cup C = \{1,3\}$ and $B \cup D = \{2,4\}$, so $(A \cup C) \times (B \cup D) = \{(1,2), (1,4), (3,2), (3,4)\}$ and this does not equal $(A \times B) \cup (C \times D)$. Therefore, the result of Exercise 2 cannot be strengthened to an equality.

3. <u>MAPPINGS</u>.

Let X and Y be nonempty sets, possibly identical. A <u>mapping</u> from X to Y is a rule which associates with each element of X a unique element of Y. (The word "function" is synonymous with "mapping", but I prefer not to use it because of its connotation with numbers, as in calculus.) If the rule is denoted by the letter "f", the mapping is denoted by the symbol "$f: X \longrightarrow Y$". If it is clear from the context what the sets X and Y are, then this mapping may be denoted just by

the letter "f". If $f: X \longrightarrow Y$ is a mapping, the set X is called the
<u>domain</u> of the mapping and the set Y is called the <u>codomain</u>. The
element of Y that is associated with the element x of X is denoted, as
usual, by "f(x)" and is called the "image of x".

If the domain and the codomain of a mapping are subsets of the
real numbers, the mapping can be pictured by a graph in the usual way.
Another useful way of picturing mappings is explained in the appendix
at the end of this section. You should look at this appendix before
continuing with this section. You will find pictures to be helpful in
understanding the concepts and the theorems relating to mappings and
I urge you to use them whenever possible.

Two mappings $f: X \longrightarrow Y$ and $g: Z \longrightarrow W$ are defined to be equal if
$X = Z$, $Y = W$, and $f(x) = g(x)$ for every x in X. If $f: X \longrightarrow Y$ is a
mapping and A is a subset of X, the <u>restriction</u> of f to A is the
mapping $g: A \longrightarrow Y$, where $g(x) = f(x)$ for every x in A. The symbol
"f|A" will be used instead of "g" for the restriction of f to A. Note
that if $A \neq X$, f|A and f are not equal because their domains are
different.

Let $f: X \longrightarrow Y$ be a mapping, A be a subset of X, and B be a subset
of Y. The (<u>direct</u>) <u>image</u> of A is defined to be

$$f^{\rightarrow}(A) = \{y \in Y \mid (\exists x)(x \in A \text{ and } y = f(x))\},$$

and the <u>inverse</u> <u>image</u> of B is defined to be

$$f^{\leftarrow}(B) = \{x \in X \mid f(x) \in B\}.$$

The set $f^{\rightarrow}(X)$ is called the <u>range</u> of the mapping.

<u>Comment</u>. Many authors write "f(A)" instead of "$f^{\rightarrow}(A)$" and "$f^{-1}(B)$"
instead of "$f^{\leftarrow}(B)$". The use of "f(A)" usually causes no confusion, so
in the future I will use it rather than the "$f^{\rightarrow}(A)$". It is customary
to call this the "image" of A, but to distinguish it from the inverse
image, the phrase "direct image" is sometimes used. I will not use the

symbol "$f^{-1}(B)$" for the inverse image of B because I believe that it does cause confusion.

Example. Since these ideas are probably new to you, I will give some examples. Let $f:R \longrightarrow R$ be the mapping given by $f(x) = x^2$ for every x in R. If $A = \{-2, 2, 3\}$, then $f(A) = \{4, 9\}$, that is, $f(A)$ is just the set of all the images of elements of A. Note that 4 comes from both -2 and 2. If $A' = \{2, 3\}$, then $f(A') = f(A) = \{4, 9\}$. If $B = \{-5, 1, 4\}$, then $f^{\leftarrow}(B) = \{-1, 1, -2, 2\}$, that is, it is the set of all real numbers whose square is -5, 1, or 4. Note that -5 is not the square of any real number, so it contributed nothing to the inverse image. If $B' = \{1, 4\}$, then $f^{\leftarrow}(B') = f^{\leftarrow}(B) = \{-1, 1, -2, 2\}$.

Comment. It is important to notice the difference between working with the direct image and working with the inverse image. If you are trying to show that some y in Y is in $f(A)$, you have to produce an x in A such that $f(x) = y$, that is, your proof must show how to find or construct such an x. However, when you are trying to show that some x in X is in $f^{\leftarrow}(B)$, you do not have to construct anything; you just look to see if $f(x)$ is in B. To make this clear, I have worked out below proofs of corresponding parts of Theorems 3.1 and 3.2. Study these proofs before you try to prove the remaining parts of these theorems.

THEOREM 3.1. Let $f:X \longrightarrow Y$ be a mapping. Then, with the obvious meanings of the symbols,

(i)* $f(\emptyset) = \emptyset$

(ii) $f(X) \subseteq Y$

(iii) $(A \subseteq A') \rightarrow (f(A) \subseteq f(A'))$

(iv) $f(\bigcup_{\gamma \in \Gamma} A_\gamma) = \bigcup_{\gamma \in \Gamma} f(A_\gamma)$ (See proof below.)

(v) $f(\bigcap_{\gamma \in \Gamma} A_\gamma) \subseteq \bigcap_{\gamma \in \Gamma} f(A_\gamma)$.

HERE AND IN THE FUTURE WHENEVER A THEOREM IS STATED, IT IS IMPLIED THAT THE CONSTRUCTION OF ITS PROOF IS AN EXERCISE.

EXERCISE 1. Construct a counterexample of a mapping to show that equality does not necessarily hold in Theorem 3.1 (v); in this example you may take $\Gamma = \{1,2\}$.

THEOREM 3.2. Let $f: X \longrightarrow Y$ be a mapping. Then, with the obvious meanings of the symbols,

(i) $f^{\leftarrow}(\emptyset) = \emptyset$

(ii) $f^{\leftarrow}(Y) = X$

(iii) $(B \subseteq B') \to (f^{\leftarrow}(B) \subseteq f^{\leftarrow}(B'))$

(iv) $f^{\leftarrow}(\bigcup_{\gamma \in \Gamma} B_\gamma) = \bigcup_{\gamma \in \Gamma} f^{\leftarrow}(B_\gamma)$ (See proof below)

(v) $f^{\leftarrow}(\bigcap_{\gamma \in \Gamma} B_\gamma) = \bigcap_{\gamma \in \Gamma} f^{\leftarrow}(B_\gamma)$

(vi)* $f^{\leftarrow}(cB) = c(f^{\leftarrow}(B))$.

EXERCISE 2. Construct a counterexample of a mapping which shows that a property analogous to Theorem 3.2 (vi) does not hold for direct images.

Proof of Theorem 3.1 (iv).

(i) Given: $y \in f(\bigcup_{\gamma \in \Gamma} A_\gamma)$

Prove: $y \in \bigcup_{\gamma \in \Gamma} f(A_\gamma)$

Proof: 1. $y \in f(\bigcup_{\gamma \in \Gamma} A_\gamma)$ (Given)

2. $(\exists x)(x \in \bigcup_{\gamma \in \Gamma} A_\gamma$ and $y = f(x))$
                                                         (Def. of image)

3. $(\exists x)(\exists \gamma)(x \in A_\gamma$ and $y = f(x))$
                                                         (Def. of union)

4. $(\exists \gamma)(\exists x)(x \in A_\gamma$ and $y = f(x))$ (See below.)

5. $(\exists \gamma)(y \in f(A_\gamma))$ (Def. of image)

6. $y \in \bigcup_{\gamma \in \Gamma} f(A_\gamma)$ (Def. of union)

(ii) Given: $y \in \bigcup_{\gamma \in \Gamma} f(A_\gamma)$

Prove: $y \in f(\bigcup_{\gamma \in \Gamma} A_\gamma)$

Proof: Reverse the steps in the preceding proof. []

Comment. The difference between lines 3 and 4 is that the two "∃" quantifiers have been reversed. To prove that such an interchange is legitimate would require a more careful study of logic than I have given in Chapter 0, but it can be done. In any event it seems to be a reasonable rule that most people accept without question. There is a similar rule that allows you to interchange two "∀" quantifiers. The situation, however, is different if there is one "∃" quantifier and one "∀" quantifier. Let $P(x, y)$ be an expression which becomes a statement when values from some universal set are substituted for x and y. The statements "$(\exists x)(\forall y)P(x, y)$" and "$(\forall y)(\exists x)P(x, y)$" do not mean the same thing. The first says that there is some x which "works" for all the y, but the second allows each y to have a different x. It should then be clear that you can go from the first to the second, but not from the second to the first. In symbols, $((\exists x)(\forall y)P(x, y)) \rightarrow ((\forall y)(\exists x)P(x, y))$, but the converse is not necessarily true. Keep this in mind when you are trying to prove Theorem 3.1 (v).

Proof of Theorem 3.2 (iv).

(i) Given: $x \in f^{\leftarrow}(\bigcup_{\gamma \in \Gamma} B_\gamma)$

Prove: $x \in \bigcup_{\gamma \in \Gamma} f^{\leftarrow}(B_\gamma)$

Proof: 1. $x \in f^{\leftarrow}(\cup_{\gamma \in \Gamma} B_\gamma)$ (Given)

2. $f(x) \in \cup_{\gamma \in \gamma} B_\gamma$ (Def. of inverse image)

3. $(\exists \gamma)(f(x) \in B_\gamma)$ (Def. of union)

4. $(\exists \gamma)(x \in f^{\leftarrow}(B_\gamma))$ (Def. of inverse image)

5. $x \in \cup_{\gamma \in \Gamma} f^{\leftarrow}(B_\gamma)$ (Def. of union)

(ii) Given: $x \in \cup_{\gamma \in \Gamma} f^{\leftarrow}(B_\gamma)$

Proof: $x \in f^{\leftarrow}(\cup_{\gamma \in \Gamma} B_\gamma)$

Proof: Reverse the steps in the preceding proof. []

Comment. Now compare the proofs of the two theorems. Note that the proof of Theorem 3.2 (iv) is simpler because you do not have an existence quantifier in Line 2 as you did have in the other proof. As a result there were not two quantifiers in any line in the second proof. Keep in mind this difference between proofs involving direct images and proofs involving inverse images. Now you should try to prove the remaining parts of Theorem 3.1 and Theorem 3.2 and do the two exercises.

THEOREM 3.3. Let $f: X \longrightarrow Y$ be a mapping. Then

(i)* $f(f^{\leftarrow}(B)) \subseteq B$

(ii) $f^{\leftarrow}(f(A)) \supseteq A$.

EXERCISE 3. Give counterexamples to show that the inclusions in Theorem 3.3 cannot, in general, be replaced by equalities.

Let X, Y, and Z be nonempty sets and $f: X \longrightarrow Y$ and $g: Y \longrightarrow Z$ be mappings. The composite mapping $gf: X \longrightarrow Z$ is defined by the rule: $(gf)(x) = g(f(x))$, for every x in X.

Comment. In analysis the composite mapping is usually denoted by "g ∘ f" so that it will not be confused with the product of two mappings. In this book there will be no occasion to multiply two mappings, so I use the simpler notation "gf" for the composite mapping. Note that in order to construct a composite mapping gf, the codemain of f must equal the domain of g. Of course, the order of the mappings in the composite is important; gf and fg, even if they can both be defined, need not be equal.

THEOREM 3.4. Let $f: X \longrightarrow Y$ and $g: Y \longrightarrow Z$ be mappings. If C is a subset of Z, then $(gf)^{\leftarrow}(C) = f^{\leftarrow}(g^{\leftarrow}(C))$.

APPENDIX

Many mappings can be pictured by graphs, such as those you used in algebra and calculus, but there is another way of illustrating mappings that is more useful for our purposes. Let me start with an example.

EXAMPLE. Let $f: R \longrightarrow R$ be defined by $f(x) = x^2$. Draw two number lines, one above the other. (See Figure 1. It is not necessary to always have the zeroes lined up or to use the same scale on both lines.)

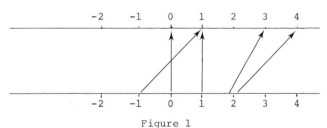

Figure 1

At each point on the lower line, say at the point x, draw an arrow whose origin or tail is at that point and whose head is on the upper line at the point x^2. Thus, arrows go from 0 to 0, from 1 to 1, from 2 to 4, from -1 to 1, etc. The mapping f can be thought of as the set

of all such arrows. Clearly, I cannot draw them all, so you should
imagine the mapping as a "cloud" of arrows starting on the lower line
and going to the upper line.

This is the scheme for illustrating mappings. Picture the
domain as some set, like a line, and the codomain as another set above
the first. A mapping is then a set of arrows that connect each point
of the lower set to its corresponding point on the upper. If, for
example, the mapping had the real plane as the domain and the real
line as the codomain, the picture would look like Figure 2.

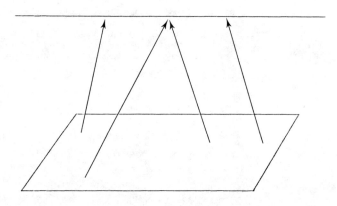

Figure 2

Usually only a few typical arrows are drawn, but if the domain
and codomain are small finite sets, all the arrows can perhaps be
drawn. This is illustrated in the next example.

EXAMPLE. Let $X = \{a,b,c,d\}$ and $Y = \{1,2,3\}$. Suppose $f: X \longrightarrow Y$
is defined by

$$f(a) = 1, \quad f(b) = 2, \quad f(c) = 2, \quad f(d) = 1.$$

Then f can be pictured as shown in Figure 3.

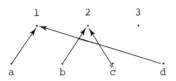

Figure 3

The various concepts of this section and later ones can be nicely illustrated by this process. I will do a few to get you started.

Let $f: X \longrightarrow Y$ be a mapping and $A \subseteq X$ and $B \subseteq Y$ be subsets. To picture the image $f(A)$, take A to be a subset of X, the lower set in the picture. Then $f(A)$ is the set of all the heads of arrows whose tails are in A. Figure 4 shows this for $f: R \longrightarrow R$, where $f(x) = x^2$ and A is the closed interval $[-1,2]$. In this case $f(A) = [0,4]$.

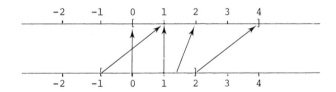

Figure 4

To picture the inverse image $f^{\leftarrow}(B)$ start by taking B as a subset of Y, the upper set. Then $f^{\leftarrow}(B)$ is the set of the tails of all the arrows whose heads lie in B. Figure 5 illustrates this for the squaring mapping with $B = [1,4]$. In this case $f^{\leftarrow}(B) = [-2,-1] \cup [1,2]$.

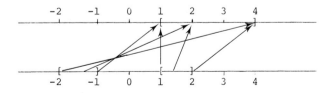

Figure 5

You should now try to illustrate the results of the theorems of this section.

You can also picture products of mappings with this method. Suppose $f: X \longrightarrow Y$ and $g: Y \longrightarrow Z$ are mappings. Draw three sets, one above the other, for X, Y and Z. (See Figure 6.) The mapping f consists of arrows going from X to Y and the mapping g of arrows going from Y to Z. The arrows for the product gf run from X to Z and are obtained by combining the arrows of f and g, that is, start with a point x in X, draw the arrow from x to f(x) in Y and then the arrow from f(x) to g(f(x)) in Z. The arrow for gf just goes directly from x to g(f(x)).

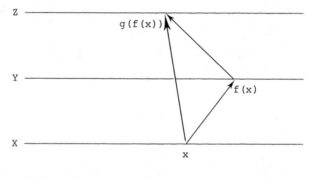

Figure 6

You should use this method of illustrating mappings in the rest of this Chapter and in the later chapters, particularly in Chapter III. Remember that the domain and codomain do not always have to be lines; sometimes small sets are better. Do not hesitate to use pictures when you are trying to construct a proof or when you are presenting your results. Even though pictures are not part of the proof, they can suggest what should be done and they can make proofs more understandable.

4. SURJECTIVE AND INJECTIVE MAPPINGS.

A mapping $f:X \longrightarrow Y$ is said to be <u>surjective</u> or <u>onto</u> Y or to be a <u>surjection</u> if $f(X) = Y$, that is, if the range of the mapping equals its codomain. To prove that $f:X \longrightarrow Y$ is surjective, you must start with an arbitrary element y in the codomain Y and then find an x in the domain X such that $f(x) = y$.

<u>Examples</u>. Let $f: \underline{Z} \longrightarrow \underline{Z}$, where $\underline{Z}$ is the set of integers, be defined by $f(x) = x + 1$. To prove that f is surjective, start with y in the codomain $\underline{Z}$ and find a value of x (there could be more than one) in the domain $\underline{Z}$ for which $f(x) = y$, that is, for which $x + 1 = y$. Obviously, $y - 1$ will work, since $f(y - 1) = (y - 1) + 1 = y$. Be careful that you start with an <u>arbitrary</u> element in the codomain. If $g: \underline{Z} \longrightarrow \underline{Z}$ is defined by $g(x) = 2x$, then g is not surjective because the range is the set of even integers and the codomain is the set of all integers. If you try a proof like the preceding one, you would start with y in $\underline{Z}$ and solve the equation $2x = y$. The solution is $x = y/2$, but $y/2$ is not necessarily an integer, that is, x is not in the domain. A common error is to start with an even integer y and conclude that the mapping is surjective. The mistake is that an even number is not an <u>arbitrary</u> element of the codomain. Of course, to prove that a mapping is not surjective, you just have to produce one element of the codomain that is not in the range.

<u>EXERCISE</u> 1. Give some other examples of mappings which are surjective and some other examples of mappings which are not surjective.

<u>THEOREM</u> 4.1. Let $f:X \longrightarrow Y$ be a mapping. Then

(i)* $f(f^{\leftarrow}(B)) = B$ for every subset B of Y iff f is surjective;

(ii) $c(f(A)) \subseteq f(cA)$ for every subset A of X iff f is surjective.

(Hint. In (i) to prove that f is surjective, consider the given condition in the case where B = Y. For the converse recall Theorem 3.3. In (ii) to prove that f is surjective, consider the given condition in the case where A = X.)

EXERCISE 2. Give an example of a surjection for which the inclusion in (ii) cannot be replaced by equality.

THEOREM 4.2.* If $f:X \longrightarrow Y$, $g:Y \longrightarrow Z$ and $h:Y \longrightarrow Z$ are mappings such that f is a surjection and gf = hf, then g = h.

(Hint. The mappings g and h are given as having the same domain and the same codomain. To prove that they are equal, you must prove that g(y) = h(y) for every y in Y.)

EXERCISE 3. Construct a counterexample to show that if the condition that f is a surjection is omitted, then the statement of the theorem need no longer be true.

Let X be a nonempty set. The <u>identity mapping</u> on X is the mapping $i_X : X \longrightarrow X$ defined by $i_X(x) = x$ for every x in X. If it is clear what set is being considered, the symbol "i" may be used in place of "i_X".

THEOREM 4.3. If $f:X \longrightarrow Y$ is a mapping, then $fi_X = f$ and $i_Y f = f$.

THEOREM 4.4. If $f:X \longrightarrow Y$ and $g:Y \longrightarrow X$ are mappings such that $fg = i_Y$, then f is a surjection.

A mapping $f:X \longrightarrow Y$ is said to be <u>injective</u> or <u>one-to-one</u> or to be an <u>injection</u> if, for all x and x' in X, the following statement is true: $(f(x) = f(x')) \rightarrow (x = x')$. This is logically equivalent to the contrapositive: $(x \neq x') \rightarrow (f(x) \neq f(x'))$. This means that f is injective iff distinct elements of X have distinct images in Y, that is, there are no duplicate images.

Examples. Let $g: \underline{Z} \longrightarrow \underline{Z}$, where $\underline{Z}$ is the set of integers, be defined by $g(x) = 2x$. To prove that g is injective, start with $g(x) = g(x')$, that is $2x = 2x'$. It follows that $x = x'$, so g is injective. The mapping $h: \underline{Z} \longrightarrow \underline{Z}$ defined by $h(x) = x^2$ is not injective because there are duplicate images, for example, $h(2)$ and $h(-2)$ both equal 4. To prove that a mapping is not injective, it suffices to find two different elements of the domain which have the same image, like the 2 and -2 above.

EXERCISE 4. Give some other examples of mappings which are injective and some which are not.

EXERCISE 5. Injectiveness and surjectiveness of mappings are independent concepts, that is, a mapping can be one without being the other. Give examples of mappings from R to R which are (i) injective, but not surjective, (ii) surjective, but not injective, (iii) both surjective and injective, and (iv) neither surjective nor injective.

THEOREM 4.5. Let $f: X \longrightarrow Y$ be a mapping. Then

   (i) $f^{\leftarrow}(f(A)) = A$ for every subset A of X iff f is injective;

   (ii)* $f(A \cap A') = f(A) \cap f(A')$ for all subsets A and A' of X iff f is injective;

   (iii)   let f be a surjection; then $c(f(A)) = f(cA)$ for every subset A of X iff f is also an injection.

(Hint. In proving that f is injective consider the given condition in the case where the subsets are singletons, that is, sets with one element.)

THEOREM 4.6. If $g: X \longrightarrow Y$, $h: X \longrightarrow Y$ and $f: Y \longrightarrow Z$ are mappings such that f is an injection and $fg = fh$, then $g = h$.

EXERCISE 6. Construct a counterexample to show that if the condition that f is an injection is omitted, then the statement of the theorem need no longer be true.

THEOREM 4.7.* If $f:X \longrightarrow Y$ and $g:Y \longrightarrow X$ are mappings such that $gf = i_X$, then f is an injection.

THEOREM 4.8. Let $f:X \longrightarrow Y$ and $g:Y \longrightarrow Z$ be mappings. Then
   (i) if f and g are surjections, then gf is surjective;
   (ii) if f and g are injections, then gf is injective.

If A is a nonempty subset of X, the <u>embedding</u> <u>mapping</u> from A to X is the mapping $e_A : A \longrightarrow X$ defined by $e_A(x) = x$ for every x in A.

EXERCISE 7. Prove that the embedding mapping $e_A : A \longrightarrow X$ is injective. Under what condition will it be surjective?

5. BIJECTIVE MAPPINGS AND INVERSES.

A mapping $f:X \longrightarrow Y$ which is both surjective and injective is called <u>bijective</u> or a <u>bijection</u> or a <u>one-to-one</u> <u>correspondence</u>. If If $f:X \longrightarrow Y$ is a bijection, then a mapping $f^{-1}:Y \longrightarrow X$ can be defined by the rule that $f^{-1}(y) = x$ iff $f(x) = y$. The next theorem shows that this is, indeed, a mapping. The mapping $f^{-1}:Y \longrightarrow X$ is called the <u>inverse</u> of the mapping $f:X \longrightarrow Y$.

THEOREM 5.1.* If $f:X \longrightarrow Y$ is a bijection, then $f^{-1}:Y \longrightarrow X$ is a mapping.

(Hint. You must show that f^{-1} has Y as domain and X as codomain and that only one value in X is associated with each element of Y. Point out where the hypothesis that f is bijective is used.)

EXAMPLE. The mapping $f:\underline{Z} \longrightarrow \underline{Z}$, defined by $f(x) = x + 1$ is easily seen to be a bijection. Let $f(x) = y$, so that $y = x + 1$. To find the inverse mapping $f^{-1}:\underline{Z} \longrightarrow \underline{Z}$, you have to solve this equation for x, that is, $x = y - 1$. Then $f^{-1}(y) = y - 1$. This illustrates the defining condition for the inverse, namely, that $f^{-1}(y) = x$ iff $f(x) = y$. Of course, the use of the letters "x" and "y" is immaterial; other letters could be used. In fact, in a case like this where the domain and the codomain are the same set, the inverse is often written with the letter "x", that is, $f^{-1}(x) = x - 1$. Such usage ordinarily does not cause confusion, but if it does for you, don't use it.

THEOREM 5.2.* If $f: X \longrightarrow Y$ is a bijection, then $f^{-1}f = i_X$ and $ff^{-1} = i_Y$.

THEOREM 5.3. If $f: X \longrightarrow Y$ and $g: Y \longrightarrow X$ are mappings such that $gf = i_X$ and $fg = i_Y$, then f is a bijection and $g = f^{-1}$.

(Hint. Use theorems of Section 4 and Theorem 5.2.)

COROLLARY 5.4. If f is a bijection, then f^{-1} is a bijection and $(f^{-1})^{-1} = f$.

Comment. Theorem 5.3 is often used to prove that a mapping $f: X \longrightarrow Y$ is a bijection. You merely try to find a mapping $g: Y \longrightarrow X$ for which $gf = i_X$ and $fg = i_Y$. Sometimes it is easy to guess what the mapping g should be.

THEOREM 5.5. If $f: X \longrightarrow Y$ and $g: Y \longrightarrow Z$ are bijections, then gf is a bijection and $(gf)^{-1} = f^{-1}g^{-1}$.

(Hint. Look at the preceding comment.)

Chapter II:
Metric Spaces

In this chapter you will begin the study of metric spaces, the main topic of this book. A metric space is essentially a set in which it is possible to speak of the distance between any two of its elements. (Mathematicians usually refer to the set as a "space" and to its elements as "points" to emphasize the geometrical aspect of this study.) The theory of metric spaces is the general theory which underlies real analysis (calculus), complex analysis, multidimensional calculus and many other subjects.

Not all the topics taken up in this chapter are of equal importance. It would be better, perhaps, to omit everything after Section 7 so as to leave more time for the remaining chapters. If you do so, it will be necessary to omit a few minor results in these later chapters. I will not point out which results should be skipped, but you should have no trouble recognizing them.

1. DEFINITION OF METRIC SPACE AND SOME EXAMPLES.

Let X be a nonempty set. A <u>metric</u> or a <u>distance</u> on X is a mapping $d: X \times X \longrightarrow R$ (the set of real numbers) such that

(M1) $d(x,y) \geq 0$ for every x and y in X,

(M2) $d(x,y) = 0$ iff $x = y$,

(M3) $d(x,y) = d(y,x)$ for every x and y in X,

(M4) $d(x,y) + d(y,z) \geq d(x,z)$ for every x, y, and z in X.

(The condition (M4) is called the "triangle inequality" for reasons that will be clear after you look at Example 2.)

A <u>metric space</u> (X,d) is a nonempty set X on which there is defined a metric d. If it is clear what the metric is, then the metric space (X,d) is often referred to simply as "the metric space X". Two metric spaces (X,d) and (Y,d') are equal if X = Y and d = d'. The following are some examples of metric spaces.

EXAMPLE 1. For the set R define d by $d(x,y) = |x - y|$. This is called the <u>usual metric</u> on R and the phrase "the metric space R" will always refer to this metric space.

EXERCISE 1. Prove that (R,d) is a metric space, that is, that d satisfies properties (M1) through (M4).

EXAMPLE 2. For R^2 = RxR take the distance between p = (x_1,y_1) and q = (x_2,y_2) to be $d(p,q) = \sqrt{(x_1 - x_2)^2 + (y_1 - y_2)^2}$. As you know from elementary analytic geometry, this is just the usual distance between points in the plane, so d is called the <u>usual metric</u> on R^2. It is easy to prove that d is a metric. Properties (M1), (M2) and (M3) are obviously true and (M4) follows from the theorem that the sum of the lengths of two sides of a triangle is greater than or equal to the length of the third side. (Equality holds when the triangle is degenerate, that is, the three vertices lie on a line.) It is because of this theorem that (M4) is called the "triangle inequality".

Some mathematicians seem to have a prejudice against using the results of high school geometry (or more technically, Euclidean geometry) in proofs. This is due to the fact that many so-called "proofs" in the early history of calculus were based on pictures or graphs. The development of the more subtle ideas of calculus showed that graphs do not always accurately represent these ideas, so more complicated proofs had to be constructed. That (R^2,d) is a metric space can be proved without using Euclidean geometry, but rather by using the properties of real numbers. If you want to see how this is done, you can look in books on advanced calculus or real analysis.

EXAMPLE 3. For $p = (x_1,y_1)$ and $q = (x_2,y_2)$ in R^2 define $d'(p,q) = \max [|x_1 - x_2|, |y_1 - y_2|]$, where max [a,b] is the larger of a or b. Then (R^2,d') is a metric space. Although the sets in Examples 2 and 3 are the same, the metrics are different, so (R^2,d) and (R^2,d') are different metric spaces.

EXERCISE 2.* Prove that (R^2,d') is a metric space. (Hint. Prove that max [a,b] + max [c,d] $\geq$ max [a + c, b + d].)

EXAMPLE 4. Let X be any nonempty set and define d by

$$d(x,y) = \begin{cases} 0, & \text{if } x = y \\ 1, & \text{if } x \neq y. \end{cases}$$

This is called the _discrete metric_ on X and (X,d) is called a _discrete metric space_. Discrete metric spaces are particularly useful when you are looking for counterexamples.

EXERCISE 3. Prove that the discrete metric is a metric.

EXAMPLE 5. (You may skip this example, particularly if you have not had much experience with least upper bounds of sets of real numbers. I have given a short explanation of least upper bounds and greatest lower bounds in an appendix to this section. You should study that appendix even if you do not examine this example. This example does show that the concept of a metric space has more extensive application than you might have suspected.)

Let X be the set of all bounded real valued functions defined on the closed interval $I = \{x \in R \mid 0 \leq x \leq 1\}$. (A real valued function f defined on I is bounded if there exists a real number K such that $|f(x)| \leq K$ for every x in I.) Define d by $d(f,g) = \sup\{|f(x) - g(x)| \mid x \in I\}$, where f and g are in X and "sup" is the supremum or least upper bound of the set.

EXERCISE 4.* Prove that (X,d) is a metric space.

THEOREM 1.1. If d is a metric on X and x, y, and z are any elements of X, then $|d(x,z) - d(y,z)| \leq d(x,y)$.

(Hint. Use the definition of absolute value to write the inequality as two inequalities without absolute values.)

APPENDIX

This is only a brief explanation of greatest lower bounds and least upper bounds; more complete discussions can be found in books on advanced calculus or real analysis.

In the following X will be a nonempty set of real numbers. A <u>lower bound</u> of X is a real number t such that $t \leq x$ for every x in X and an <u>upper bound</u> of X is a real number u such that $u \geq x$ for every x in X. Not every set of real numbers has a lower bound or an upper bound. For example, the set of all positive integers has no upper bound, but any real number less than or equal to 1 is a lower bound. The set X has no lower bound iff for every real number b there exists an x in X such that $x < b$. A similar statement can be made for a set with no upper bound.

A <u>greatest lower bound</u> or <u>infinum</u> of X is the largest of the lower bounds of X, if there is such a number. The infinum of X is written as "inf X". More precisely, inf X is a real number s such that (1) s is a lower bound of X and (2) if t is any lower bound of

X, then $t \leq s$. Statement (2) can also be written as follows: if $t' > s$, then t' is not a lower bound of X. Similarly, a <u>least upper bound</u> or <u>supremum</u> of X, written "sup X", is the smallest of the upper bounds of X, if there is such a number. More precisely, $v = \sup X$ if (1) v is an upper bound of X and (2) if u is any upper bound of X, then $u \geq v$. Statement (2) can also be written as follows: if $u' < v$, then u' is not an upper bound of X.

A fundamental property of the system of real numbers states that if a set X of real numbers has a lower bound, then it has a greatest lower bound and if it has an upper bound, then it has a least upper bound. The proof of this property cannot be given here because it would have to be based on a detailed investigation of the real numbers, including their construction from the set of rational numbers. Nevertheless, we will use this property freely. It is important that you understand that to show that a set X of real numbers has an infinum, you only have to show that it has a lower bound. Once you have proved that it has a lower bound, you can apply this property to say that it has an infinum. A similar remark holds for suprema.

<u>EXERCISE</u> 5. Prove that if X and Y are nonempty sets of real numbers which have lower bounds and if $X \subseteq Y$, then $\inf X \geq \inf Y$. State and prove the corresponding result for suprema.

2. <u>CLOSED AND OPEN BALLS; SPHERES</u>.

The elements of a metric space are usually called the <u>points</u> of the metric space. Given a point p of a metric space (X,d) and a positive real number r, the <u>open</u> ball, <u>closed</u> ball, and <u>sphere</u> with <u>center</u> p and <u>radius</u> r are, respectively, the sets

$B(p;r) = \{x \in X \mid d(p,x) < r\}$,

$B^*(p;r) = \{x \in X \mid d(p,x) \leq r\}$,

$S(p;r) = \{x \in X \mid d(p,x) = r\}$.

In situations where there are several metric spaces under consideration, the above symbols are modified by indicating the metric space by a subscript, as in "$B_X(p;r)$".

EXAMPLE. (I will use the standard symbols for intervals of real numbers, that is, (a,b) is the open interval $\{x \in R \mid a < x < b\}$ and $[a,b]$ is the closed interval $\{x \in R \mid a \leq x \leq b\}$. The parenthesis indicates that the number is not included in the interval and the bracket indicates that it is included. For example $(a,b] = \{x \in R \mid a < x \leq b\}$.) In the metric space R (Example 1.1) $B(p;r)$ is the open interval $(p - r, p + r)$ and $B^*(p;r)$ is the closed interval $[p - r, p + r]$. The sphere $S(p;r)$ consists just of the two points, $p - r$ and $p + r$.

EXERCISE 1. Describe the open balls, closed balls and spheres of the metric spaces in the other examples of Section 1. Draw pictures of each for a typical point and radius.

THEOREM 2.1. If $r > r' > 0$, then $B(p;r) \supseteq B(p;r')$.

EXERCISE 2. Give an example of a metric space in which $B(p;r)$ might equal $B(p;r')$ even though $r > r' > 0$.

THEOREM 2.2.* If p and q are distinct points of a metric space, there exist two open balls, one with center at p and the other at q, that are disjoint, that is, their intersection is empty.

(Hint. To see how to determine the radii, draw a picture for the metric space R^2 (Example 1.2). Your proof, of course, cannot use the picture because that applies only to this special space.)

3. **OPEN SETS.**

A set G in the metric space X is called an <u>open set</u> if for every point p of G there exists a positive real number r such that the open ball $B(p;r)$ is contained in G. The real number r depends on the point p, so different values of r might have to be chosen for different points of G.

<u>THEOREM</u> 3.1. The empty set and the whole space X are open sets.

<u>THEOREM</u> 3.2.* An open ball is an open set.

(Hint. Draw a picture of $B(p;r)$ for the metric space R^2 (Example 1.2). If $x \in B(p;r)$, you must find an s such that $B(x;s) \subseteq B(p;r)$. Of course, your proof cannot use the picture.)

<u>EXERCISE</u> 1. For the metric space R prove that an "open" interval is open and that a singleton, that is, a one point set, is not open.

<u>EXERCISE</u> 2. What are the open sets in a discrete metric space? Give a proof of your answer.

<u>THEOREM</u> 3.3. Let $(G_\gamma \mid \gamma \in \Gamma)$ be a nonempty family of open sets of the metric space X. Then

    (i) $\bigcup_{\gamma \in \Gamma} G_\gamma$ is an open set,

    (ii)   if Γ is finite, $\bigcap_{\gamma \in \Gamma} G_\gamma$ is an open set.

<u>EXERCISE</u> 3. Give an example of an intersection of infinitely many open sets which is not open. You might consider intervals of R.

<u>THEOREM</u> 3.4.* A subset of the metric space X is open iff it is the union of a family of open balls.

(Hint. If G is open, for every p in G there exists an open ball $B(p;r_p)$ contained in G. (The radius is written as "r_p" to remind you that it depends on the point p.) Consider the family of these open balls for all p in G.)

EXERCISE 4. Give examples of open sets which are not open balls for the metric spaces of Examples 1.1, 1.2, and 1.3.

Comment. Open sets are among the most important objects in geometry and analysis. As you go through this chapter notice how often new ideas are defined in terms of open sets or that there are theorems which give necessary and sufficient conditions for these new ideas in terms of open sets.

4. CLOSED SETS.

Let A be a subset of the metric space X. A point p of X is called an accumulation point of A (or sometimes a limit point of A or a cluster point of A) if every open ball with center at p contains a point of A other than p itself (if p happens to be a point of A), that is, if for every positive real number r, $(B(p;r) - \{p\}) \cap A \neq \emptyset$. A subset of a metric space is said to be closed if it contains all its accumulation points.

EXAMPLE. In the metric space R let A be the set of all positive real numbers. The number 0 is not in A, but 0 is an accumulation point of A. To see this take any open ball with center at 0, say, $B(0;r)$. Then $B(0;r)$ is the open interval $(-r, r)$, so it contains points of A, for example, $r/2$. Therefore, 0 is an accumulation point of A. A positive number, like 2, is also an accumulation point of A. An open ball with center at 2 and radius r is the open interval $(2 - r, 2 + r)$ and this contains points of A other than 2, for example, $2 + r/2$. (Of course, it might contain negative numbers, but this does not invalidate the condition.) On the other hand, a negative number, like -2, is not an accumulation point of A. To show this, you just have to produce one open ball with center at -2 that contains no positive numbers, for example, (-3, -1). In the general case, if p is a negative number, take q = -p and use the open ball (p - q, p + q).

THEOREM 4.1.* A subset of a metric space is closed iff its complement is open.

THEOREM 4.2. The empty set and the whole space X are closed sets. Closed balls are closed sets.

(Hint. To prove this theorem use Theorem 4.1. In the last part it might be helpful to draw a picture of a closed ball in R^2.)

EXERCISE 1. Prove that a "closed interval" in the metric space R is closed. What are the closed sets of a discrete metric space?

THEOREM 4.3. Let $(F_\gamma \mid \gamma \in \Gamma)$ be a nonempty family of closed sets of a metric space. Then

(i) $\cap_{\gamma \in \Gamma} F_\gamma$ is a closed set,

(ii) if Γ is finite, $\cup_{\gamma \in \Gamma} F_\gamma$ is a closed set.

(Hint. Use Theorems 3.3 and 4.1.)

EXERCISE 2. Show that the finiteness of Γ in Theorem 4.3 (ii) is essential by giving an example of an infinite family of closed sets whose union is not closed.

COROLLARY 4.4. A sphere S(p;r) is a closed set.

(Hint. Express S(p;r) in terms of closed and open balls.)

THEOREM 4.5.* If p is an accumulation point of a subset A of a metric space X, then for any positive real number r, B(p;r) contains infinitely many distinct points of A.

(Hint. Assume B(p;r) contains only finitely many points of A.)

COROLLARY 4.6. Any finite subset of a metric space is closed. In particular, every singleton, that is, one-point subset, is closed.

5. **CLOSURE OF A SET.**

The closure of a subset A of a metric space X is defined to be the union of A and the set of accumulation points of A. The closure of A is denoted by "$\bar{A}$" or by "$Cl_X(A)$". The latter symbol is used particularly when there are several metric spaces under consideration.

EXERCISE 1. Give examples of closures for the various metric spaces in the examples of Section 1. Show that in R the closure of an open interval (a,b) is the closed interval [a,b].

THEOREM 5.1. A subset A of the metric space X is closed iff $\bar{A} = A$. In particular, $\bar{\emptyset} = \emptyset$ and $\bar{X} = X$.

Comment. Of course, $A \subseteq \bar{A}$, so if you use Theorem 5.1 to prove that a set A is closed, you only have to prove that $\bar{A} \subseteq A$.

THEOREM 5.2.* A point p is an element of $\bar{A}$ iff for every positive real number r, $B(p;r) \cap A \neq \emptyset$.

THEOREM 5.3. Let A and B be subsets of the metric space X. Then
(i)   if $A \subseteq B$, then $\bar{A} \subseteq \bar{B}$;
(ii)* $\bar{A}$ is closed;
(iii)   if F is a closed subset of X such that $A \subseteq F$, then $\bar{A} \subseteq F$;
(iv)* $\bar{A}$ is the intersection of all the closed subsets of X which contain A;
(v) $\bar{A} \cup \bar{B} = \overline{A \cup B}$ and $\bar{A} \cap \bar{B} \supseteq \overline{A \cap B}$.

(Hint. To prove (ii) show that if $p \in Cl_X(\bar{A})$, then $p \in \bar{A}$. To do this use the condition of Theorem 5.2.)

EXERCISE 2. Give an example to show that equality need not hold in the second part of Theorem 5.3 (v).

Comment. Theorem 5.3 (iv) gives another way of characterizing the closure of a subset A of a metric space X, namely, $\overline{A}$ is the smallest closed set which contains A. By the "smallest" closed set containing A, I mean that $\overline{A}$ is a closed set containing A and that if F is any closed set containing A, then F must also contain $\overline{A}$. To see that this does characterize the closure, let A' be a closed set containing A having the property that if F is any closed set containing A, then F contains A'. I have to show that A' = $\overline{A}$. Since A' is a closed set containing A, Theorem 5.3 (iii) gives $\overline{A} \subseteq$ A'. On the other hand, since $\overline{A}$ is a closed set containing A, the defining condition of A' gives A' $\subseteq \overline{A}$. Therefore, A' = $\overline{A}$.

6. <u>DIAMETER OF A SET</u>; <u>BOUNDED SETS</u>.

The <u>diameter</u> of a nonempty subset A of a metric space X is defined to be $\delta(A) = \sup \{d(x,y) \mid x \in A \text{ and } y \in A\}$, if this supremum exists, and to be infinity if it does not. ("δ" is the lower case Greek letter "delta".) <u>A bounded set</u> is a nonempty subset of X whose diameter is not infinity. To prove that a subset A of X is bounded, it is only necessary to prove that $\{d(x,y) \mid x \in A \text{ and } y \in A\}$ has an upper bound. (See the appendix to Section 1.) An <u>unbounded set</u> is a nonempty set that is not bounded. This means that if A is an unbounded set, then for any positive real number K, however large, there exist points p and q in A such that $d(p,q) > K$.

EXAMPLES. I want to prove that the diameter of the open interval (0,1) in the metric space R is 1. First observe that if x and y are in (0,1), then $d(x,y) < 1$, so 1 is an upper bound of $\{d(x,y) \mid x \in (0,1) \text{ and } y \in (0,1)\}$. (Incidentally, this proves that (0,1) is bounded.) Next let r be any positive real number less than 1. I will show that there exist points p and q in (0,1) for which $d(p,q) > r$. This will prove that no number less than 1 is an upper bound for $\{d(x,y) \mid x \in (0,1) \text{ and } y \in (0,1)\}$, which means that $\delta((0,1)) = 1$.

Take s to be a real number such that $r < s < 1$ and let $p = (1 - s)/2$ and $q = (1 + s)/2$. Since $0 < s < 1$, it is easily seen that p and q are in $(0,1)$. Finally, $d(p,q) = |(1 - s)/2 - (1 + s)/2| = |-s| = s > r$. Similar arguments will show that $\delta([0,1]) = 1$ and that, in general, $\delta((a,b)) = \delta([a,b]) = b - a$. You should observe that if $\delta(A) = r$, it is not necessarily true that there are two points p and q in A such that $d(p,q) = r$. The example of $(0,1)$ shows this. The set of positive integers is an unbounded subset of R. To prove this, let K be any positive real number. Take x to be a positive integer greater than $K + 1$ and y to be 1. Then $d(x,y) > |(K + 1) - 1| = K$.

THEOREM 6.1. Let A and B be nonempty subsets of a metric space X. Then (i) $A \subseteq B$ implies $\delta(A) \leq \delta(B)$;

(ii) $\delta(A) = 0$ iff A is a singleton;

(iii) $\delta(B(p;r)) \leq 2r$ and $\delta(B^*(p;r)) \leq 2r$.

(Hint. In (i) do not forget the case where the sets are unbounded. In (iii) remember that the supremum of a set is less than or equal to any upper bound.

EXERCISE. Give examples to show that equality need not hold in Theorem 6.1 (iii).

LEMMA 6.2. If A and B are bounded subsets of a metric space X and if $a \in A$ and $b \in B$, then for any x and y in $A \cup B$, $d(x,y) \leq d(a,b) + \delta(A) + \delta(B)$.

(Hint. Take special cases: x and y both in A; both in B; one in A and the other in B.)

THEOREM 6.3. The union of two bounded subsets of a metric space is bounded. The union of finitely many bounded subsets of a metric space is bounded.

(Hint. The proof of the second statement of the theorem should be done by mathematical induction. If you are not familiar with this method of proof, you will find a short explanation in Appendix M at the end of the book.)

THEOREM 6.4.* A nonempty subset of a metric space X is bounded iff there is a point p in X such that A is contained in some closed ball with center at p.

7. SUBSPACES OF A METRIC SPACE.

Let (X,d) be a metric space and Y be a nonempty subset of X. It is obvious that the restriction of d to YxY is a metric on Y. (Properties (M1) through (M4) hold for all the points of the larger set X, so they also hold for the points of the subset Y.) The restriction of d to YxY will again be denoted by "d", and Y will be called a subspace of X when this restriction is the metric that is used on Y.

EXAMPLE. The set Q of all the rational numbers can be taken as a subspace of the metric space R. This just means that if x and y are rational numbers, the distance between x and y is $|x - y|$, just as it is in R. On the other hand, if Q is given the discrete metric, then it would not be a subspace of R because a different metric is being used on Q.

If Y is a subspace of X, it is important to distinguish between the metric spaces X and Y. For example, if $p \in Y$, the open balls $B_X(p;r)$ and $B_Y(p;r)$ are not necessarily the same. $B_X(p;r) = \{x \in X \mid d(x,p) < r\}$, but $B_Y(p;r) = \{x \in Y \mid d(x,p) < r\}$. The latter contains only points of Y.

LEMMA 7.1. If Y is a subspace of the metric space X and if p is a point of Y, $B_Y(p;r) = B_X(p;r) \cap Y$.

If A is a subset of the subspace Y of the metric space X, then you must distinguish between the statements, "A is open in the metric space Y" and "A is open in the metric space X". (See Exercise 1, below.) Usually, we just say, "A is open in Y", for the former statement and, "A is open in X", for the latter. Similar care must be taken for closed sets and for all other concepts of metric spaces.

THEOREM 7.2.* Let Y be a subspace of the metric space X and $A \subseteq Y$. Then A is open in Y iff there exists a set G which is open in X such that $A = G \cap Y$.

(Hint. Use Theorem 3.4.)

EXERCISE 1. Give an example of a subspace of R in which there is an open subset for the subspace which is not open in R.

THEOREM 7.3. Let Y be a subspace of the metric space X. Then every subset of Y which is open in Y is also open in X iff Y is open in X.

THEOREM 7.4.* Let Y be a subspace of the metric space X and $A \subseteq Y$. Then A is closed in Y iff there exists a set F which is closed in X such that $A = F \cap Y$.

(Hint. Use Theorem 7.2 and Theorem 4.1. Be careful to take complements relative to the space in question.)

EXERCISE 2. State and prove the theorem for closed sets that corresponds to Theorem 7.3.

EXERCISE 3. Let Y be a subspace of the metric space X and $A \subseteq Y$. State and prove a theorem which relates the closure of A in Y to the closure of A in X.

8. INTERIOR OF A SET.

Let A be a subset of the metric space X. A point p of A is called an *interior point* of A if there exists a positive real number r such that $B(p;r) \subseteq A$. The set all the interior points of A is called the *interior* of A and is denoted by "A°" or by "$\text{Int}_X(A)$". Of course, $A° \subseteq A$.

EXERCISE 1. Give some examples of interiors of subsets of the metric spaces in the examples of Section 1.

THEOREM 8.1.* If A is a subset of the metric space X, then

    (i) $\overline{A} = c(\text{Int}_X(cA))$,

    (ii) $A° = c(\text{Cl}_X(cA))$.

Comment. Although the properties of interiors can all be proved on the basis of the definition, Theorem 8.1 gives a way of proving these properties by using the known results about closures. I will give an example of this process and I suggest you use it to prove the theorems of this section.

EXAMPLE. (This is a proof of Theorem 8.3 (ii) below.) I will prove that A° is open. By Theorem 8.1, $A° = c(\overline{cA})$. But $\overline{cA}$ is closed since it is a closure of a set, so $c(\overline{cA})$ is open.

THEOREM 8.2. A subset A of the metric space X is open iff $A° = A$. In particular, $\emptyset° = \emptyset$ and $X° = X$.

THEOREM 8.3. Let A and B be subsets of the metric space X. Then
(i)   if $A \subseteq B$, then $A° \subseteq B°$;
(ii)   A° is open;
(iii)   if G is an open subset of X such that $A \supseteq G$, then $A° \supseteq G$;
(iv)   A° is the union of all the open subsets of X which are contained in A;

(v) $A° \cap B° = (A \cap B)°$ and $A° \cup B° \subseteq (A \cup B)°$.

EXERCISE 2. Give an example to show that equality need not hold in the second part of Theorem 8.3 (v).

EXERCISE 3. State and prove a result about the interior of a set that corresponds to the result about closures that is discussed in the comment at the end of Section 5.

9. BOUNDARY OF A SET.

Let A be a subset of the metric space X. A point p of X is called a <u>boundary point</u> of A if for every positive real number r, $B(p;r) \cap A \neq \emptyset$ and $B(p;r) \cap (cA) \neq \emptyset$, that is, every open ball with center at p contains points in both A and cA. The set of all the boundary points of A is called the <u>boundary</u> of A and is denoted by "Bd(A)".

EXERCISE 1. Show that in R the boundary of the interval [a,b) is the set {a,b}. What is the boundary of a singleton in R? What is the boundary of B(p;r) in a discrete metric space?

THEOREM 9.1. Let A be a subset of a metric space. Then $Bd(A) = \overline{A} \cap (\overline{cA})$.

COROLLARY 9.2. Let A be a subset of a metric space. Then
(i)   Bd(A) is a closed set;
(ii)   Bd(A) = Bd(cA);
(iii) $Bd(\overline{A}) \subseteq Bd(A)$;
(iv) $Bd(A°) \subseteq Bd(A)$.

(Hint. Use Theorem 9.1.)

EXERCISE 2. Give examples in R to show that equality need not hold in Corollary 9.2 (iii) and (iv).

THEOREM 9.3. Let A be a subset of a metric space. Then
(i)*   A is open iff A ∩ Bd(A) = ∅;
(ii)   A is closed iff Bd(A) ⊆ A.

THEOREM 9.4. Let A be a subset of a metric space. Then
(i)*   A° = A - Bd(A);
(ii) $\overline{A}$ = A ∪ Bd(A).

Comment. These two theorems are not particularly useful in proving other results, but they should give you a "feeling" for open and closed sets, interiors and closures. Theorem 9.3 (i) shows that a set is open iff it contains none of its boundary points and (ii) shows that it is closed iff it contains all of its boundary points. Similarly, Theorem 9.4 (i) shows that the interior of a set is obtained by discarding from the set all its boundary points, while (ii) shows that the closure is obtained by taking the set and adjoining to it all its boundary points.

10. <u>DENSE SETS</u>.

Let A be a subset of the metric space X. A is said to be <u>dense</u> in X if $\overline{A}$ = X. Of course, it is always true that $\overline{A}$ ⊆ X, since X is the universe; so to prove that a subset A of the metric space X is dense in X, it suffices to prove that X ⊆ $\overline{A}$. Therefore, you would start with an arbitrary point of X and prove that it is in $\overline{A}$.

EXERCISE. Outline an argument to show that the set of rational numbers is dense in R. A complete proof of this result would require a detailed examination of the real number system, so I am only asking for an informal explanation based on the intuitive properties of the number line.

THEOREM 10.1.* A subset A of the metric space X is dense in X iff the only closed set containing A is X.

THEOREM 10.2. A subset A of the metric space X is dense in X iff A intersects every nonempty open set of X.

THEOREM 10.3. A subset of the metric space X is dense in X iff the complement of A has an empty interior, that is, $(cA)° = \emptyset$.

11. AFTERWORD.

In this chapter you have studied many of the elementary "geometric" properties of metric spaces. The concept of an open set is particularly important because most of the other concepts can be expressed in terms of open sets. You probably have heard about topology. A <u>topological space</u> is a nonempty set in which there is a reasonable concept of open sets. In particular, the properties in Theorems 3.1 and 3.3 must hold. In this sense, a metric space is a topological space, although it has special properties which some topological spaces do not possess. In the material which follows this chapter you will notice that I often express concepts, theorems and proofs in terms of open sets. When I do this, I am using the fact that the concept, theorem or proof is really topological. If you ever take a topology course, you will see that many of these ideas and proofs carry over to topological spaces without change.

Chapter III:
Mappings of Metric Spaces

In the last chapter you learned about some of the "geometry" of metric spaces. These geometrical ideas are more general than those you studied in high school, such as the congruence or similarity of triangles. They underlie all of geometry and are fundamental for the study of analysis, the branch of mathematics that develops and extends the ideas of calculus. In calculus you studied derivatives and integrals, which in turn were based on the idea of limits. It is possible to define limits in a metric space, but I will examine a more basic concept, that of continuous mappings of metric spaces.

Generally speaking, whenever mathematicians have some mathematical structure, like a metric space, they want to see how two of the objects of that structure are related. They do this by studying mappings between these objects, but not just any mappings, rather those that are closely related to the structure being examined. In the theory of metric spaces these are the continuous mappings. In this chapter you will learn precisely what such mappings are and what are their elementary properties.

As in Chapter II not everything in this chapter must be covered. I would suggest that you omit Section 2 if time is short or if you did not study Sections 8 and 10 of Chapter II.

1. CONTINUOUS MAPPINGS.

Let (X,d) and (Y,d') be metric spaces and f be a mapping from X to Y. The mapping f is said to be _continuous at the point_ x_o of x if for every positive real number ε there exists a positive real number δ such that if x is any point of X which satisfies the condition $d(x,x_o) < \delta$, then $d'(f(x),f(x_o)) < \varepsilon$. ("$\varepsilon$" and "$\delta$" are the lower case Greek letters epsilon and delta.) A mapping $f: X \longrightarrow Y$ which is continuous at every point of X is said to be _continuous_.

Comment. If you look up the meaning of continuous function in a calculus book, you will probably find a definition which is given in terms of limits, but usually there is also an "ε, δ - definition" derived from this. You will see without trouble that this definition is exactly the same as the definition of continuity at a point x_o given above, when it is applied to the case where the two metric spaces both equal R.

THEOREM 1.1.* The mapping $f: X \longrightarrow Y$ is continuous at x_o iff for every open ball $B_Y(f(x_o); \varepsilon)$ in Y there exists an open ball $B_X(x_o; \delta)$ in X such that $f(B_X(x_o; \delta)) \subseteq B_Y(f(x_o); \varepsilon)$.

EXERCISE 1. Let X be a discrete metric space and Y be any metric space. Prove that any mapping $f: X \longrightarrow Y$ is continuous.

EXERCISE 2. Let X and Y be metric spaces and $f: X \longrightarrow Y$ be a constant mapping, that is, $f(x) = b$, for a fixed element b in Y. Prove that f is continuous.

EXERCISE 3. Let $i_X: X \longrightarrow X$ be the identity mapping on the metric space X. Prove that i_X is continuous.

EXERCISE 4. Let $f: R \longrightarrow R$ be defined by $f(x) = -1$, if $x < 0$, and $f(x) = 1$, if $x \geq 0$. Prove that f is not continuous at 0.

EXERCISE 5. Theorem 1.1 gives a necessary and sufficient condition for a mapping to be continuous at a point. Take the negation of this condition to write a careful statement for a necessary and sufficient condition that a mapping is not continuous at a point. Compare this statement with the way that you worked Exercise 4.

THEOREM 1.2.* Let f be a mapping from the metric space X to the metric space Y. Then f is continuous iff for every open set G in Y, $f^{\leftarrow}(G)$ is an open set in X.

Comment. This is a most important theorem. It gives a characterization of continuity that is not expressed in terms of continuity at each point. In most cases you will find it easier to use this characterization of continuity rather than ε, δ - arguments. (See also the Afterword to Chapter II.)

THEOREM 1.3. Let f be a mapping from the metric space X to the metric space Y. Then f is continuous iff for every closed set F in Y, $f^{\leftarrow}(F)$ is a closed set in X.

(Hint. Use the property that the complement of a closed set is open.)

THEOREM 1.4. Let f be a mapping from the metric space X to the metric space Y. Then f is continuous iff for every subset A of X, $f(\bar{A}) \subseteq \overline{f(A)}$.

(Hint. To prove that the condition holds when f is continuous, take y in $f(\bar{A})$ and show that $B_Y(y;\varepsilon) \cap f(A) \neq \emptyset$ for every $\varepsilon > 0$. To prove the converse use Theorem 1.3.)

EXERCISE 6. Give an example to show that equality need not hold in the condition of Theorem 1.4.

THEOREM 1.5. Let X, Y and Z be metric spaces and $f: X \longrightarrow Y$ and $g: Y \longrightarrow Z$ be continuous mappings. Then $gf: X \longrightarrow Z$ is continuous.

EXERCISE 7. Show by an example that if $f: R \longrightarrow R$ is continuous, the image of an open set need not be open.

2.   CONTINUOUS MAPPINGS AND SUBSPACES.

Let X be a metric space and A be a nonempty subset of X. Recall that the embedding mapping from A to X is the mapping $e_A: A \longrightarrow X$ defined by $e_A(x) = x$ for every x in A. (See Section I.4.) The embedding mapping is an injection.

THEOREM 2.1. If A is given the subspace metric, the embedding mapping $e_A: A \longrightarrow X$ is continuous.

Let $f: X \longrightarrow Y$ be a mapping of metric spaces and let A be a nonempty subset of X. Recall that the restriction of f to A is the mapping $f|A: A \longrightarrow Y$ defined by $(f|A)(x) = f(x)$ for every x in A. When there are statements made about the continuity of $f|A$, it is to be understood that the subspace metric is being used on A.

THEOREM 2.2. Let X and Y be metric spaces, A be a nonempty subset of X, and $f: X \longrightarrow Y$ be a mapping. Then
(i) $f|A = fe_A$;
(ii)   if f is continuous, then $f|A$ is continuous.

LEMMA 2.3. Let X and Y be metric spaces, A be a nonempty subset of X, and $f: X \longrightarrow Y$ be a mapping. Then $f|A$ is continuous at the point x_0 of A iff for every positive real number ε there exists a positive real number δ such that $f(B_X(x_0; \delta) \cap A) \subseteq B_Y(f(x_0); \varepsilon)$.

THEOREM 2.4.* Let X and Y be metric spaces, A be a nonempty subset of X, and $f: X \longrightarrow Y$ be a mapping. If $f|A$ is continuous, then f is continuous at every interior point of A.

EXERCISE. Give an example of a mapping $f:R \longrightarrow R$ to show that in Theorem 2.4 the mapping f might not be continuous at every point of A, even though $f|A$ is continuous.

THEOREM 2.5.* Let f and g be continous mappings from the metric space X to the metric space Y. If A is a dense subset of X and if $f|A = g|A$, then $f = g$.

(Hint. Assume there is a point x_0 for which $f(x_0) \neq g(x_0)$. Use Theorem II 2.2.)

Comment. Theorem 2.5 shows that a continuous mapping is completely determined by its values on a dense subset of its domain. For example, in high school you learned the meaning of 2^x for every rational number x. By Theorem 2.5 there can be at most one continuous function $f(x) = 2^x$ for all real numbers x, that is, there can be only one way to define 2^x for real numbers in such a way that the function is continuous. Be careful! I have not said that there is such a continuous real function; I have only said that there can be no more than one. The problem of extending a mapping from a dense subset of the domain to the entire domain is a difficult one and cannot always be done so that the result is continuous. In the case of 2^x it can be done, but I will not show that here.

3. UNIFORM CONTINUITY.

Let (X,d) and (Y,d') be metric spaces and f be a mapping from X to Y. The mapping f is said to be <u>uniformly continuous</u> on X if for every positive real number ε there exists a positive real number δ such that for any pair of points x_1 and x_2 in X for which $d(x_1,x_2) < \delta$, then $d'(f(x_1),f(x_2)) < \varepsilon$.

Comment. When you are proving that a mapping $f:X \longrightarrow Y$ is continuous at a point x_0, the δ that you must find usually depends on both ε and

x_0. If the mapping is uniformly continuous, the δ depends on ε alone. This distinction might appear to be minor, but it is extremely important in many situations.

THEOREM 3.1. A uniformly continuous mapping $f: X \longrightarrow Y$ is continuous.

EXERCISE 1.* Let X be the interval $(0,1]$ with the subspace metric of R. Define the mapping $f: X \longrightarrow R$ by $f(x) = 1/x$. Prove that f is continuous at each point of X, but f is not uniformly continuous on X. (The point of this exercise is to show that not every continuous mapping is uniformly continuous, while Theorem 3.1 shows that the converse is true. The proof is tricky and I suggest that you do not spend too much time on it. If you are not successful in working it out, look up my proof in the appendix.)

THEOREM 3.2. Let X, Y and Z be metric spaces and $f: X \longrightarrow Y$ and $g: Y \longrightarrow Z$ be uniformly continuous mappings. Then the mapping $gf: X \longrightarrow Z$ is uniformly continuous.

Comment. Notice that your proof of Theorem 3.2 used ε and δ, rather that just open sets. This is not an accident. I mentioned at the end of Chapter II that a metric space has some properties that a topological space does not have. This means that some results in the theory of metric spaces cannot be expressed just in terms of open sets. Uniform continuity is one of these.

THEOREM 3.3. The identity mapping on a metric space is uniformly continuous.

The last theorem leads to a generalization. If (X,d) and (Y,d') are metric spaces and $f: X \longrightarrow Y$ is a surjection, then f is called an isometry if for every pair of points x_1 and x_2 in X, $d(x_1, x_2) = d'(f(x_1), f(x_2))$. If there is an isometry from a metric space to a metric space, the spaces are called isometric.

THEOREM 3.4. Let $f: X \longrightarrow Y$ be an isometry. Then
(i)   f is injective (and hence bijective);
(ii)   f is uniformly continuous.

Comment. An isometry $f: X \longrightarrow Y$ sets up a one-to-one correspondence between the points of X and the points of Y in such a way that the distance between a pair of points of X equals the distance between the corresponding points of Y. This means that the metric spaces X and Y are "essentially" the same, that is, as metric spaces they have the same properties. They might differ in other respects; one might, for example, have some algebraic property that the other does not have, but any property that can be phrased in terms of metric concepts that holds for one space holds for the other also. Usually we regard two isometric spaces as being the same and do not distinguish between them.

EXERCISE 2. Let X and Y both be the set of real numbers, but let X have the discrete metric and Y have the usual metric. Prove that the identity mapping $i_X : X \longrightarrow Y$ is a uniformly continuous bijection, but is not an isometry.

Comment. Exercise 2 shows that the converse of Theorem 3.4 is not true.

Chapter IV:
Sequences in Metric Spaces

Sequences of real numbers and their limits play a large role in calculus and in applied mathematics. In this chapter you will study sequences in the broader context of metric spaces, but the results, of course, apply to sequences of real numbers. The chapter begins with a careful study of sequences in general and then passes to metric spaces where the concept of limit can be introduced. This is followed by a look at how sequences are related to metric properties, like closure or continuity. Finally, there is a brief study of Cauchy sequences and complete metric spaces.

If there is not enough time to study the entire chapter or if you want to look into some of the topics in the other chapters, you could omit Sections 4 and 5. However, if you do omit them, I suggest that you return to them when you do have time. The material in these sections is not difficult and it is important in analysis courses.

1. SEQUENCES.

A sequence in a set X is a mapping from the set $\underline{N}$ of positive integers to X. Instead of using the mapping notation, mathematicians usually use a subscript notation for sequences. Thus, a sequence is a mapping $a: \underline{N} \longrightarrow X$ and the value of the mapping at the positive integer i is denoted by "a_i", rather than by "a(i)". The sequence itself is denoted by the symbol "(a_n)", rather than by "$a: \underline{N} \longrightarrow X$". (The "n" here is just part of the symbol.) It is essential that you distinguish between "(a_n)", which denotes the sequence, and "a_n", which denotes the value of the sequence at n. In spite of the possibility for misunderstanding, this notation is used by most mathematicians, and it will probably cause you no trouble after a little while.

Different sequences are denoted by different letters, for example, by "(a_n)" or by "(x_n)". Two sequences are equal if they are equal as mappings, that is, $(a_n) = (b_n)$ iff they are both sequences in the same set X and for every positive integer n, $a_n = b_n$. It is important to note that a sequence need not be an injective mapping, so it is possible that $a_i = a_j$ even though i and j are not equal. You must distinguish between a sequence and its range. For example, let (a_n) be the sequence in R that is defined by the rule: $a_n = 1$, if n is odd, and $a_n = 2$, if n is even. The range of (a_n) is the set $\{1,2\}$. If (b_n) is the sequence in R that is defined by the rule: $b_n = 2$, if n is odd, and $b_n = 1$, if n is even, then the range of (b_n) is also $\{1,2\}$, but $(a_n) \neq (b_n)$.

When you think of a sequence, you probably should imagine a listing of elements of X. For example, the sequence (a_n) of the last paragraph is the listing 1, 2, 1, 2, 1, ..., and the sequence (b_n) is the listing 2, 1, 2, 1, 2, ... Any such listing clearly defines a sequence, if you presume that you know what occurs where the dots are written.

A sequence (b_n) is a <u>subsequence</u> of the sequence (a_n) if there exists an increasing mapping $\phi: \underline{N} \longrightarrow \underline{N}$ such that for every n in $\underline{N}$, $b_n = a_{\phi(n)}$. (To say that ϕ is an increasing mapping means that if i and j are positive integers with $i > j$, then $\phi(i) > \phi(j)$.) Roughly speaking, this means that (b_n) is a listing of some of the elements of (a_n) and the relative order of the terms in the listing is unchanged. For example, if (a_n) is the sequence given above and if $\phi(x) = 2x$, then $(a_{\phi(n)})$ is a sequence consisting entirely of the number 2. In fact, $b_1 = a_{\phi(1)} = a_2 = 2$, $b_2 = a_{\phi(2)} = a_4 = 2$, etc. Often a subsequence of (a_n) is denoted by "(a_{i_n})", but I will write "$(a_{\phi(n)})$". The only difference in these notations is that in the first one the increasing mapping is denoted by "i" and the subscript notation is used, while in the second the mapping notation is used. Since a mapping from $\underline{N}$ to $\underline{N}$ is a sequence, the first notation is consistent with the sequence notation, but I have found that it usually causes confusion among students. Therefore, I will continue to use the ϕ notation. Whenever I refer to a subsequence, it is presumed that the mapping ϕ, or whatever other letter is used, is an increasing mapping from $\underline{N}$ to $\underline{N}$.

<u>EXERCISE 1</u>. Prove that any sequence is a subsequence of itself.

<u>EXERCISE 2</u>.* Prove that if (b_n) is a subsequence of (a_n) and (c_n) is a subsequence of (b_n), then (c_n) is a subsequence of (a_n).

<u>EXERCISE 3</u>. Write an explicit formula for the mapping ϕ that gives the subsequence that is obtained by omitting the first m terms of a sequence (a_n).

LEMMA 1.1.* Let $\phi: \underline{N} \longrightarrow \underline{N}$ be an increasing mapping. Then for every positive integer n, $\phi(n) \geq n$.

(Hint. Use mathematical induction. There is a discussion of mathematical induction in an appendix at the end of the book.)

Often a sequence is constructed by recursion. In such a process the first term, a_1, is defined by some rule and then it is explained how to define a_{n+1} when the preceding terms, $a_1, a_2, \ldots, a_n$ are assumed to be known. Usually the elements of the sequence have to satisfy some condition, which might depend on the subscripts. You have to show that a_1 satisfies this condition and so does a_{n+1} when you assume that $a_1, a_2, \ldots, a_n$ all satisfy it.

EXAMPLE. A geometric sequence in R with a given ratio r can be defined by recursion. Assume that a_1 is any real number. If $a_1, a_2, \ldots, a_n$ are assumed to be known, define a_{n+1} to be ra_n.

It can be proved by using mathematical induction that the process of recursion actually does define a sequence, but the logical basis of such "definition by recursion" is a little too complicated to go into here. I will assume that recursion is a legitimate process for defining sequences and will use it freely.

EXERCISE 4.* Let X be an infinite subset of the positive integers. Prove that there exists a sequence of distinct points of X by using recursion to construct such a sequence. Try to be as explicit as possible in the recursion process. (Hint. The set of positive integers has the property that every nonempty subset has a smallest element.) Next take X to be an infinite subset of any set and prove that there exists a sequence of distinct points of X by the use of recursion. In this case your proof cannot be as explicit as the preceding proof, but you must show that, in principle, the recursion process can be carried out.

2. <u>SEQUENCES</u> <u>IN</u> <u>METRIC</u> <u>SPACES</u>.

Let (x_n) be a sequence of points of a metric space X. A point p of X is called a <u>limit</u> of the sequence (x_n) if for every positive real number ε there exists a positive integer N such that x_n is in $B(p;\varepsilon)$ for all $n > N$. If (x_n) has a limit p, we say that the sequence (x_n) is <u>convergent</u> or that (x_n) <u>converges</u> to p and we write "$\lim x_n = p$". A sequence which is not convergent is said to <u>diverge</u> or to be <u>divergent</u>.

EXERCISE 1. Let (x_n) be a constant sequence in a metric space X, that is, there exists a point p of X such that $x_n = p$ for every positive integer n. Prove that (x_n) is convergent.

EXERCISE 2. Let p and q be distinct points of a metric space X and let (x_n) be the sequence in which $x_n = p$, if n is odd, and $x_n = q$, if n is even. Prove that (x_n) is divergent.

(Hint. Assume that the sequence converges to b. Take the two cases, $b = p$ and $b \neq p$, and derive contradictions, using Theorem II 2.2.)

THEOREM 2.1. If (x_n) is a convergent sequence in a metric space X, then the limit of (x_n) is unique.

(Hint. Use Theorem II 2.2.)

THEOREM 2.2. If $(x_{\phi(n)})$ is a subsequence of the convergent sequence (x_n), then $(x_{\phi(n)})$ is convergent and converges to $\lim x_n$.

EXERCISE 3. Give an example of a divergent sequence in R that has a convergent subsequence.

THEOREM 2.3.* Let A be a subset of a metric space X. Then a point p of X is an accumulation point of A iff there exists a sequence (x_n) of points A, none of which equals p, such that $\lim x_n = p$.

(Hint. Construct a sequence (x_n) such that for each positive integer n, $x_n \in B(p;1/n)$.)

COROLLARY 2.4. Let A be a subset of a metric space X. If (x_n) is a sequence of points of A which converges to the point p of X, then $p \in \bar{A}$.

COROLLARY 2.5. A subset A of a metric space X is closed iff every convergent sequence of points of A has its limit in A.

(Hint. If A is not closed, then there is an accumulation point of A which is not in A.)

EXERCISE 4. Give an example of a subset A of R which is not closed and a sequence of points of A whose limit is not in A.

THEOREM 2.6. Let X and Y be metric spaces and $f: X \longrightarrow Y$ be a mapping which is continuous at the point p of X. If (x_n) is a sequence of points of X that converges to p, then $(f(x_n))$ is a sequence of points of Y that converges to the point $f(p)$.

3. CLUSTER POINTS OF A SEQUENCE.

A point p of a metric space X is called a **cluster point** of the sequence (x_n) of points of X (or sometimes a **limit point** of the sequence or an **accumulation point** of the sequence) if for every positive real number ε and every positive integer N there exists an integer $n > N$ such that $x_n \in B(p;\varepsilon)$. This clearly means that for every open ball with center at p there are infinitely many values of the subscript n such that x_n is in that ball.

THEOREM 3.1. The limit of a convergent sequence is a cluster point of the sequence.

EXERCISE 1. Give an example of a sequence in R that has a cluster point which is not the limit of the sequence.

THEOREM 3.2.* A point p of a metric space X is a cluster point of the sequence (x_n) iff there is a subsequence $(x_{\phi(n)})$ of the sequence that converges to p.

(Hint. Use recursion to define a subsequence in such a way that $x_{\phi(n)} \in B(p;1/n)$ for every n.)

COROLLARY 3.3. If a sequence has more than one cluster point, then it is divergent.

EXERCISE 2. Give an example of a divergent sequence that has exactly one cluster point.

THEOREM 3.4. Let (x_n) be a sequence in a metric space X and A be the range of the sequence. If p is an accumulation point of A, then p is a cluster point of (x_n).

EXERCISE 3. Give an example of a sequence in R and a point p of R which is a cluster point of the sequence, but not an accumulation point of the range of the sequence.

THEOREM 3.5. If a sequence in a metric space X has a finite range, then it has a cluster point in X.

THEOREM 3.6. Let p be a cluster point of the sequence (x_n) of points of a metric space X and $f: X \longrightarrow Y$ be a mapping of metric spaces which is continuous at p. Then f(p) is a cluster point of the sequence $(f(x_n))$.

4. CAUCHY SEQUENCES.

A sequence (x_n) of points of a metric space X is a <u>Cauchy sequence</u> if for every positive real number ε there exists a positive integer N such that for all integers m and n, each greater than N, $d(x_m, x_n) < \varepsilon$. Roughly speaking, this says that all the points of the sequence after some index value are close to each other.

<u>THEOREM</u> 4.1. A convergent sequence in a metric space is a Cauchy sequence.

<u>EXERCISE</u> 1. Write a careful statement for the negation of the Cauchy condition. Use this to prove that the sequence (x_n) in R where $x_n = n$, for every n, is not a Cauchy sequence.

<u>Comment</u>. It is a property of the real number system that the converse of Theorem 4.1 is true in R, that is, a sequence in R is convergent iff it is a Cauchy sequence. (There is a proof of this in Chapter VI.) However, there are metric spaces in which a Cauchy sequence need not converge.

<u>EXERCISE</u> 2. Let X be the interval (0,1] of R with the subspace metric and (x_n) be the sequence in X defined by $x_n = 1/n$, for every positive integer n. Show that (x_n) is a Cauchy sequence, but that it does not converge in X.

(Hint. Use the fact that (x_n) is a convergent sequence in R.)

<u>Comment</u>. You might think that this exercise is not fair since the sequence fails to converge only because the limit was expressly eliminated from the space. However, as will be explained in the next section, this is typical of the metric spaces in which Cauchy sequences need not converge.

THEOREM 4.2.* If (x_n) is a Cauchy sequence in a metric space X and if there is a cluster point p of the sequence in X, then (x_n) converges to p.

COROLLARY 4.3. If the Cauchy sequence (x_n) has a convergent subsequence, then (x_n) is convergent.

COROLLARY 4.4. If a Cauchy sequence has a finite range, then it is convergent.

THEOREM 4.5.* The range of a Cauchy sequence is a bounded set.

(Hint. Show that there is an integer N such that for all $n > N$, $x_n \in B(x_{N+1};1)$. Then take into account the points x_n with $n \leq N$ and find a suitable closed ball with center at x_{N+1} that contains all the points of the sequence.)

THEOREM 4.6. If $f:X \longrightarrow Y$ is a uniformly continuous mapping of metric spaces and if (x_n) is a Cauchy sequence in X, then $(f(x_n))$ is a Cauchy sequence in Y.

EXERCISE 3. Show that Theorem 4.6 need not be true if the condition of uniform continuity is replaced by continuity.

(Hint. See Exercise 1 of Section III.3.)

5. COMPLETE METRIC SPACES.

A metric space X is said to be complete if every Cauchy sequence of points of X converges to a point of X. In a complete space it is, therefore, possible to determine if a sequence is convergent without knowing the limit; merely show that it is a Cauchy sequence.

Comment. The metric space R is complete, but there are metric spaces which are not complete. (See Section 4.) There is an important property of metric spaces which says that if X is a metric space which is not complete, then there is a complete metric space Y which contains X as a subspace. (This property will not be proved in this book.) This means that if (x_n) is a Cauchy sequence in X, then it does converge, but the limit might be in Y rather than in X. This shows why the example of a non-complete metric space in Section 4 can be said to be a typical example. The next two theorems show that I could not have used a closed interval in that example.

THEOREM 5.1. If X is a complete metric space and Y is a closed subset of X, then Y with the subspace metric is complete.

THEOREM 5.2. If Y is a subset of the metric space X and if Y with the subspace metric is complete, then Y is closed in X.

THEOREM 5.3.* If every bounded infinite subset of a metric space X has an accumulation point in X, then X is complete.

Chapter V:
Connectedness

In this chapter you will study the concept of connectedness. Roughly speaking, the idea is to distinguish those spaces and sets that split up into several pieces from those that are all one piece. This can be done in several different ways and I will take up just one such criterion. After the general ideas have been worked out, you will look at the situation on the real line and prove an important theorem - the Intermediate Value Theorem.

This chapter and the next are independent of each other. If you prefer, you can skip this chapter completely and go to the next, but if you do study this chapter, I suggest you cover it completely. One of the results (Theorem 2.2) has a complicated proof, so I have written it out completely, rather than having you work on the proof. It would be worthwhile, however, for you to study the proof carefully, because it contains some interesting uses of least upper bounds and closed and open sets.

1. CONNECTED SPACES AND SETS.

A metric space is said to be connected if it is not the union of two nonempty disjoint open subsets. A metric space which is not connected is called disconnected. This means that a metric space X is disconnected iff there exist nonempty open subsets A and B such that A ∩ B = ∅ and A ∪ B = X. Since the definition of a connected metric space is given in negative terms, to prove that a metric space is connected, you should usually try an indirect proof, that is, assume it is disconnected and derive a contradiction. You will prove in Section 2 that R is a connected metric space.

EXERCISE 1. Prove that a discrete metric space with more than one point is disconnected.

EXERCISE 2. Let X = (0,1) ∪ [2,3] be given the subspace metric of R. Prove that X is disconnected.

THEOREM 1.1. A metric space is connected iff it is not the union of two nonempty disjoint closed subsets.

(Hint. If X = A ∪ B and A ∩ B = ∅, then A and B are complements of each other.)

You already know that if X is a metric space, ∅ and X are both open and closed. The next theorem shows that in the case of a connected metric space these are the only subsets that are both open and closed.

THEOREM 1.2.* A metric space X is connected iff the only subsets of X which are both open and closed are ∅ and X.

EXERCISE 3. Give an example of a subspace X of R which contains a subset different from ∅ and X that is both open and closed in X.

In addition to studying connected metric spaces, it is necessary to study connected subsets of a metric space, which itself might not be connected. A nonempty subset Y of a metric space X is said to be connected if it is connected when considered as a subspace of X. If the subset Y is not connected, then it is called disconnected. In

Section 2 you will prove that intervals of R are connected subsets of R.

THEOREM 1.3.* A nonempty subset Y of a metric space X is disconnected iff there exist open sets A and B in X with the following four properties: (i) $A \cap Y \neq \emptyset$, (ii) $B \cap Y \neq \emptyset$, (iii) $(A \cap B) \cap Y = \emptyset$, and (iv) $Y \subseteq A \cup B$. Furthermore, the word "open" can be replaced by the word "closed".

EXERCISE 4. Prove that the set of nonzero real numbers is a disconnected subset of R by using Theorem 1.3 with open sets and again with closed sets.

Comment. To prove that a subset Y of a metric space X is connected you should assume that there exist open (or closed) subsets A and B in X such that $Y \subseteq A \cup B$ and $(A \cap B) \cap Y = \emptyset$ and try to prove that $A \cap Y$ or $B \cap Y$ is empty.

THEOREM 1.4. If Y is a connected subset of the metric space X and if Z is a subset of X such that $Y \subseteq Z \subseteq \overline{Y}$, then Z is connected. In particular, if Y is connected, then $\overline{Y}$ is connected.

(Hint. Use the process described in the comment with closed sets.)

THEOREM 1.5.* If $(A_\gamma \mid \gamma \in \Gamma)$ is a nonempty family of connected subsets of a metric space X and if $\bigcap_{\gamma \in \Gamma} A_\gamma \neq \emptyset$, then $\bigcup_{\gamma \in \Gamma} A_\gamma$ is connected.

2. CONNECTED SETS IN R.

It is convenient to extend the concept of interval on the real line. An _interval_ of R is a subset I of R which has the property that for any two distinct points p and q of I, all the points between p and q are also in I. It follows that a subset I of R is not an interval iff there exist two points p and q in I such that some point between p and q is not in I.

EXERCISE. Show that an open interval (a,b) and a closed interval [a,b] of R are intervals in the above sense. Also show that an open ray $(a,\infty) = \{x \in R \mid x > a\}$ and a closed ray $[a,\infty) = \{x \in R \mid x \geq a\}$ are intervals. (There are also open and closed rays of the form $(-\infty,b)$ and $(-\infty,b]$. These, along with intervals of the form (a,b] and [a,b), are also intervals. The proofs are similar to those you have been asked to do.) Finally, show that R itself is an interval.

THEOREM 2.1. Any connected subset of R containing more than one point is an interval.

(Hint. Prove the contrapositive.)

A proof of the following theorem is tricky, so I have written it out.

THEOREM 2.2. A closed interval [a,b] of R is connected.

Proof: Let I = [a,b]. Suppose that A and B are nonempty disjoint subsets of I that are closed in I such that $I = A \cup B$. Since I is closed in R and A and B are closed in I, A and B are closed in R. (See Exercise 2 of Section II 7.) Since b is an upper bound of I, it is an upper bound of A. Therefore, sup A exists; call it p. For any $\varepsilon > 0$, $B_R(p;\varepsilon) \cap A \neq \emptyset$, so $p \in \bar{A} = A$. But since A is open in I, there exists δ such that $B_I(p;\delta) \subseteq A$, that is, $B_R(p;\delta) \cap I \subseteq A$. If $p \neq b$, then there exists a point q such that $q \in I$ and $p < q < p + \delta$. Then q is a point of A bigger than sup A, which is impossible. Hence, p = b, so $b \in A$. The same sort of argument will show that $b \in B$, which is a contradiction, since $A \cap B = \emptyset$. []

THEOREM 2.3. Any nonempty interval of R is connected.

(Hint. Let I be a nonempty interval and $p \in I$. For each x in I let

I_x be the closed interval whose endpoints are p and x. Use theorems 2.2 and 1.5.)

Comment. It is obvious that a singleton is connected. Actually, a singleton in R is an interval according to the above definition. (It satisfies the condition of the definition in a "vacuous" sense; the definition involves an "If ..., then ..." statement with the "if ..." part false for a singleton.) The above theorems show that a nonempty subset of R is connected iff it is an interval. In particular, R is connected.

3. MAPPINGS OF CONNECTED SPACES AND SETS.

THEOREM 3.1.* If $f:X \longrightarrow Y$ is a continuous mapping of metric spaces and X is connected, then f(X) is connected in Y.

(Hint. Let A be a nonempty subset of f(X) that is open and closed in f(X). Then there exist an open set G and a closed set F in Y such that $A = G \cap f(X) = F \cap f(X)$. Look at $f^{\leftarrow}(G)$ and $f^{\leftarrow}(F)$.)

COROLLARY 3.2. If $f:X \longrightarrow Y$ is a continuous mapping of metric spaces and if A is a connected subset of X, then f(A) is a connected subset of Y.

(Hint. Use Theorem 3.1 and Theorem III 2.2.)

THEOREM 3.3. (The Intermediate Value Theorem) Let $f:R \longrightarrow R$ be a continuous mapping and p and q be points of R with p < q and $f(p) \neq f(q)$. If t is any real number between f(p) and f(q), then there exists a real number s such that $p \leq s \leq q$ and f(s) = t.

Comment. This theorem is used extensively in numerical processes. For example, suppose $f:R \longrightarrow R$ is continuous and you want to solve the equation f(x) = 0. If you can find two real numbers p and q such that f(p) and f(q) have opposite signs, then by the Intermediate

Value Theorem you know that there is a root of the equation between p and q. To get a better approximation of the root, you merely have to find such numbers p and q that are close together.

Chapter VI:
Compactness

In a metric space there are usually sequences which do not have cluster points, but there are some metric spaces in which every sequence does have a cluster point. Such spaces are said to be sequentially compact. A related concept is that of compactness and a metric space with this property is called a compact space. As you will see, these two concepts are equivalent for metric spaces, but since they are not equivalent for the more general case of topological spaces, it is customary to study them separately. After you have learned about some of the properties of compact and sequentially compact metric spaces and have proved that they are equivalent, you will apply the results to the metric space R and derive some very important theorems of analysis.

At this stage in the study of metric spaces some of the proofs are rather complicated, so rather than having you spend a considerable amount of time trying to work them, I have written them out completely. You should examine them carefully to understand the special devices that are used. (For example, see the use of the factor 1/2 in the proof of Theorem 2.5.) These might seem to be mere tricks, and perhaps they are, but after you have seen them used several times, they cease to be tricks and become techniques. Eventually, you will have at hand a number of such proof techniques and will be able to use them in new situations.

1. **COMPACT SPACES AND SETS.**

A family $(G_\gamma \mid \gamma \in \Gamma)$ of subsets of a metric space X is called a cover of X if $X = \bigcup_{\gamma \in \Gamma} G_\gamma$. If each G_γ is an open subset of X, then the cover is called an open cover of X. If $\Gamma' \subseteq \Gamma$, then $(G_\gamma \mid \gamma \in \Gamma')$ is called a subfamily of $(G_\gamma \mid \gamma \in \Gamma)$. If a subfamily of a cover of X is still a cover of X, then it is called a subcover of X. If Γ' is finite, the subfamily is called a finite subfamily and if it is a subcover of X, it is called a finite subcover of X. Now after all these preliminaries comes the important definition. A metric space X is called compact if every open cover of X contains a finite subcover. A subset of X is called a compact set if it is compact when considered as a subspace of X.

EXAMPLE. In the metric space R for each integer n, let G_n be the open interval $(n - 1, n + 1)$. Then $(G_n \mid n \in Z)$, where Z is the set of all integers, is an open cover of R, but it contains no finite subcover. To see this latter statement, assume there is a finite subcover and let m be the largest subscript such that G_m is in the subcover. Then $m + 2$ is not in the union of the sets of the subcover. This is a contradiction. This result shows that R is not a compact metric space. In Section 4 you will see that the compact subsets of R are the closed and bounded subsets.

EXERCISE 1. Let X be a discrete metric space with infinitely many points. Prove that X is not compact.

EXERCISE 2. Let A be the half-open interval $(0,1]$ of R. Show that A is not a compact subset of R.

There are several equivalent formulations of compactness which are useful. The following results give some of them.

THEOREM 1.1. A metric space X is compact iff every family $(F_\gamma \mid \gamma \in \Gamma)$ of closed subsets of X such that $\cap_{\gamma \in \Gamma} F_\gamma = \emptyset$ contains a finite subfamily whose intersection is also empty.

A family of subsets is said to have the <u>finite intersection property</u> if every finite subfamily has a nonempty intersection.

COROLLARY 1.2. A metric space is compact iff every family of closed subsets with the finite intersection property has a nonempty intersection.

THEOREM 1.3. A subset A of a metric space X is compact iff for every family $(G_\gamma \mid \gamma \in \Gamma)$ of open sets of X for which $A \subseteq \cup_{\gamma \in \Gamma} G_\gamma$, there is a finite subfamily whose union contains A.

THEOREM 1.4.* A closed subset of a compact metric space is compact.

(Hint. Given a family of open sets whose union contains the closed subset, adjoin one more open set to the family to get an open cover of the whole space.)

LEMMA 1.5. Let A be a compact subset of the metric space X and let p be a point of X that is not in A. Then there exists an open ball B(p;r) which is disjoint from A.

(Hint. Use Theorem II 2.2 with p and with each point of A to find a family of open sets whose union contains A.)

EXERCISE 3. Show that the condition that A be compact in Lemma 1.5 is essential by giving an example of a non-compact subset A and a point p not in A such that $B(p;r) \cap A$ is not empty, regardless of the value of r.

THEOREM 1.6. A compact subset of a metric space is closed.

THEOREM 1.7. A compact subset of a metric space is bounded.

(Hint. If A is a compact subset of a metric space, take the family $(B(x;1) \mid x \in A)$.)

EXERCISE 4. The above results show that a compact subset of a metric space is closed and bounded. Use a discrete metric space to show that the converse of this statement is not true.

2. MAPPINGS OF COMPACT SPACES.

THEOREM 2.1. Let $f: X \longrightarrow Y$ be a continuous mapping of metric spaces. If X is a compact metric space, then $f(X)$ is a compact subset of Y.

COROLLARY 2.2. Let $f: X \longrightarrow Y$ be a continuous mapping of metric spaces. If A is a compact subset of X, then $f(A)$ is a compact subset of Y.

COROLLARY 2.3. Let $f: X \longrightarrow Y$ be a continuous mapping of metric spaces. If X is compact and A is a closed subset of X, then $f(A)$ is closed in Y.

EXERCISE. Give an example of a continuous mapping $f: R \longrightarrow R$ and a closed subset A of R such that $f(A)$ is not closed.

THEOREM 2.4. Let $f: X \longrightarrow Y$ be a continuous bijection of metric spaces. If X is a compact metric space, then the mapping f^{-1} is continuous.

(Hint. Prove that if $F \subseteq X$, then $(f^{-1})^{\leftarrow}(F) = f(F)$.)

THEOREM 2.5. Let $f: X \longrightarrow Y$ be a continuous mapping of metric spaces. If X is compact, then f is uniformly continuous.

Proof: Take a fixed positive ε. For each p in X, there exists $\delta(p) > 0$ such that $f[B_X(p;\delta(p))] \subseteq B_Y(f(p);\varepsilon/2)$. Take the open cover $(B_X(p;\delta(p)/2) \mid p \in X)$. Since X is compact, there exists a finite subset $\{p_1, p_2, \ldots, p_n\}$ of X such that the family $(B_X(p_i;\delta(p_i)/2) \mid 1 \leq i \leq n)$ covers X. Take δ to be the minimum of the $\delta(p_i)/2$, $1 \leq i \leq n$. I claim that this δ will satisfy the condition of uniform continuity for the original ε.

Take any points x_1 and x_2 in X such that $d(x_1,x_2) < \delta$. There exists a point p_i in the subset found above such that $x_1 \in B_X(p_i;\delta(p_i)/2)$, that is, $d(x_1,p_i) < \delta(p_i)/2$. Then, $d(x_2,p_i) \leq d(x_2,x_1) + d(x_1,p_i) < \delta + \delta(p_i)/2 < \delta(p_i)$. Therefore, x_1 and x_2 are both in $B_X(p_i;\delta(p_i))$, so $f(x_1)$ and $f(x_2)$ are both in $B_Y(p_i;\varepsilon/2)$. Then $d'(f(x_1),f(x_2)) \leq d'(f(x_1),f(p_i)) + d'(f(p_i),f(x_2)) < \varepsilon/2 + \varepsilon/2 = \varepsilon$, as required. []

Comment. After you have finished Section 4, you will recognize in Theorem 2.5 the important theorem of analysis that states that a continuous real function whose domain is a closed and bounded set of real numbers is uniformly continuous.

3. SEQUENTIAL COMPACTNESS.

A metric space X is said to be **sequentially compact** if every sequence in X has a cluster point in X. This is equivalent to saying that every sequence in X has a convergent subsequence.

EXAMPLE. Let X be the interval $(0,1]$ of R with the subspace metric. X is not sequentially compact. To see this, take the sequence (x_n), where $x_n = 1/n$ for every positive integer n. This sequence has only one cluster point in R, namely 0, and since 0 is not in X, the sequence has no cluster points in X.

THEOREM 3.1. A metric space X is sequentially compact iff every infinite subset of X has an accumulation point in X.

(Hint. If A is an infinite subset of a sequentially compact space, construct a sequence of distinct points of A and prove that a cluster point of the sequence is an accumulation point of A.)

COROLLARY 3.2. A sequentially compact metric space is complete.

THEOREM 3.3.* If X is a compact metric space, then every infinite subset of X has an accumulation point in X. Hence, every compact metric space is sequentially compact.

(Hint. Assume that A is an infinite subset with no accumulation points in X. For each point p of X find an open ball with center at p whose intersection with A is $\emptyset$ or $\{p\}$.)

A finite set of points $\{p_1, p_2, \ldots, p_n\}$ of a metric space X is called an <u>ε-net</u> if the family $(B(p_i;\varepsilon) \mid 1 \leq i \leq n)$ covers X. A metric space is called <u>totally bounded</u> or <u>precompact</u> if there is an ε-net for every positive real number ε. Thus, X is totally bounded iff for every positive real number ε, X can be covered by a finite family of open balls of radius ε.

THEOREM 3.4. If a metric space has an ε-net for some positive real number ε, then it is bounded. Hence, a totally bounded metric space is bounded.

THEOREM 3.5. A sequentially compact metric space is totally bounded. Hence, a sequentially compact metric space is bounded.

Proof: Assume X is not totally bounded, so there exists $\varepsilon > 0$ such that there is no ε-net. Define by recursion a sequence (p_n) such that for each n, $d(P_n, p_i) \geq \varepsilon$ for $1 \leq i < n$. To do this, start with any p_1 in X. Assume that $p_1, p_2, \ldots, p_n$ have been defined such that for $m = 1, 2, \ldots, n$, $d(p_m, p_i) \geq \varepsilon$, $1 \leq i < m$. Since $X \neq \bigcup_{1 \leq i \leq n} B(p_i;\varepsilon)$, it is possible to choose p_{n+1} to be a point not in this union. Then $d(p_{n+1}, p_i) \geq \varepsilon$ for $i = 1, 2, \ldots, n$, as required.

Since X is sequentially compact, (p_n) has a convergent subsequence, which is then a Cauchy sequence. For the ε used above there must exist a positive integer N such that for any pair of points of the subsequence with subscripts larger than N, say $p_{\phi(m)}$ and $p_{\phi(n)}$, $d(p_{\phi(m)}, p_{\phi(n)}) < \varepsilon$. This contradicts the condition that was imposed on (p_n), that is, if, say, $\phi(n) > \phi(m)$, then $d(p_{\phi(m)}, p_{\phi(n)}) \geq \varepsilon$. This proves that X is totally bounded. []

Comment. The following lemma shows that the sets of an open cover of a sequentially compact space overlap each other by a definite amount. Of course, this is also true for a compact space. (See Theorem 3.3.)

LEMMA 3.6. If X is a sequentially compact metric space and $(G_\gamma \mid \gamma \in \Gamma)$ is an open cover of X, then there exists a positive real number ε such that every open ball of radius ε is contained in at least one of the sets G_γ.

Proof: Assume no such ε exists. For each positive integer n there exists a point p_n of X such that $B(p_n; 1/n)$ is not contained in any G_γ. This gives a sequence (p_n); let p be a cluster point of (p_n). There exists some G_α such that $p \in G_\alpha$, and since G_α is open, there exists a positive integer m such that $B(p; 2/m) \subseteq G_\alpha$. Because p is a cluster point of (p_n), there exists $n > m$ such that $p_n \in B(p; 1/m)$. I claim that $B(p_n; 1/n) \subseteq G_\alpha$ which will be the desired contradiction in this indirect proof.

If $x \in B(p_n; 1/m)$, then $d(x,p) \leq d(x, p_n) + d(p_n, p) < 1/m + 1/m = 2/m$, so $x \in B(p; 2/m)$. Recall that $n > m$, so $1/n < 1/m$. Then,

$B(p_n;1/n) \subseteq B(p_n;1/m) \subseteq B(p;2/m) \subseteq G_\alpha$, as claimed. []

THEOREM 3.7. A sequentially compact metric space is compact.

(Hint. Given an open cover of X, take the ε of Lemma 3.6. Since the space is totally bounded, it has an ε-net for this ε. Use this to get a finite subcover.)

Comment. Theorems 3.3 and 3.7 show that compactness and sequential compactness are equivalent for metric spaces. Theorem 3.1 shows that these concepts are equivalent to the condition that every infinite subset has an accumulation point in the space.

4. COMPACT SUBSETS OF R.

THEOREM 4.1. A closed interval [a,b], where a and b are real numbers, is a compact subset of R.

Proof: Let $(G_\gamma \mid \gamma \in \Gamma)$ be a family of open sets of R such that $[a,b] \subseteq \cup_{\gamma \in \Gamma} G_\gamma$. Define $A = \{x \in [a,b] \mid [a,x]$ is contained in a union of finitely many $G_\gamma\}$. $A \neq \emptyset$ because $a \in A$. If $x_1 \in A$ and $x_2 < x_1$, then $x_2 \in A$, because $[a,x_2] \subseteq [a,x_1]$. Furthermore, if $x \in A$ and if $x < b$, then there exists $y > x$ such that $y \in A$. To see this, note that x is in some G_α so there exists $r > 0$ such that $(x - r, x + r) \subseteq G_\alpha \cap [a,b]$. Then, $[a, x + r/2] = [a,x] \cup (x - r, x + r/2]$ is contained in a union of finitely many G_γ, namely those involved with [a,x] and G_α.

Let $c = \sup A$. By the last result above, c cannot be less than b, so $c = b$. Since b is in some G_β, there exists $\varepsilon > 0$ such that $(b - \varepsilon, b + \varepsilon) \subseteq G_\beta$. Also since $b = \sup A$, $b - \varepsilon$ is not an upper bound of A. Therefore, there is some x in A such that $b - \varepsilon < x \leq b$. From the first result in the last paragraph, $b - \varepsilon \in A$. It follows, as above, that $[a,b] = [a,b-\varepsilon] \cup (b-\varepsilon, b]$ is contained in a union of finitely many G_γ. This proves that $[a,b]$ is compact. []

THEOREM 4.2. A subset of R is compact iff it is closed and bounded.

(Hint. A bounded subset of R is contained in some closed interval.)

Comment. When applied to R, Theorem 3.1 is called the Bolzano-Weierstrass Theorem and Theorem 4.2 is called the Heine-Borel Theorem.

THEOREM 4.3. Let X be a compact metric space and $f: X \longrightarrow R$ be a continuous mapping. Then there exist points p and q in X such that $f(p) = \inf f(X)$ and $f(q) = \sup f(X)$.

EXERCISE. Show that the condition that X be compact in this theorem is essential by constructing a counterexample in the case where X is not compact.

THEOREM 4.4. The metric space R is complete.

(Hint. See Theorem IV 5.3.)

Afterword

Perhaps you have noticed that mathematics books usually do not end, they merely stop. This is not because the authors are poor stylists; rather it is in the nature of the subject. Mathematics is constantly growing. Even in those parts that you might think have been completed long ago, new results are still being discovered. So it is impossible to end a mathematics book; the subject itself has not ended. An author has to decide if he has accomplished what he set out to do and then stop.

In this book I intended to help you learn to construct proofs and to teach you some of the elementary theory of metric spaces. I hope I have been successful, but before I stop, let me go over these two goals with you.

Think back to your attitude toward mathematical proofs before you started this book. Perhaps you found them mysterious or maybe you were even afraid of them. I would like to think that your attitude has changed for the better. I hope that you have confidence in your ability to construct, understand, and criticize proofs. You should know that when you are faced with a proof you have to recognize what is given and what you have to prove. You should have learned some logic and have become familiar with enough mathematical techniques so that in many cases you can actually construct the proof. Of course, you have not learned everything there is to know about proving theorems. No one knows it all, because, like mathematics, proof theory itself is growing. For example, there have been some recent advances in the use of computers in very long proofs. However, you should have learned enough to carry you through most of college mathematics.

In your future work in mathematics you will further develop your ability to construct proofs. I have several suggestions to make. What should you do when you come across a proof in your mathematical reading? First, it is not always necessary to study the proof in detail; to do so in every case would be too time consuming. However, you cannot ignore proofs. To understand mathematics, you must see how everything hangs together. Usually this means that you have to see how the results are proved. This does not mean that you must understand every detail, so do not be afraid to skip details, at least on a first reading. Secondly, you should not immediately begin to read a proof. Stop and think about it first. Ask yourself what is given and what has to be proved. Give a little thought about how you might actually prove it. If you have time, try to write out a proof. If you do not have the time, then read the proof. If your idea was not used, try to see why it was not. Most importantly, try to understand the general reasoning pattern in the proof. Thirdly, if a proof involves some unfamiliar process, study it carefully. It might, as I said at the beginning of Chapter VI, be a trick that will develop eventually into a useful technique. Notice that the reading of mathematics is not a passive occupation. You must be willing to take an active part in it.

My second goal was to teach you some of the elementary theory of metric spaces. I tried to cover those topics that would be useful in a real analysis course or a beginning topology course. Don't stop here. If you have an opportunity to take a course in general topology, I suggest you do so. If you do not, then get a textbook in topology and start working through it on your own. With the background you now have, you should find most of the material understandable. I am sure that you will find it an interesting subject. I encourage you, in any event, to begin to read mathematics. You do not have to like everything; people react differently to the various areas of mathematics. Try to find a topic that interests you and then go into it as deeply as you can. Remember that you do not have to be a professional mathematician to enjoy mathematics.

I hope that I have accomplished what I set out to do in this book and I hope you enjoyed using it. Goodbye!

Appendix M:
Mathematical Induction

There is an important proof technique in mathematics called "mathematical induction", although a better name would be "proof by recursion". Let (P_n) be a sequence of statements, that is, for each positive integer n, P_n is a statement, either true or false. The process of mathematical induction gives a way of proving that all these statements are true by proving two results - one is a special case of the statements in the sequence and the other is a general result linking any two successive statements in the sequence.

PRINCIPLE OF MATHEMATICAL INDUCTION. Let (P_n) be a sequence of statements. Then all the statements in the sequence are true if the following two conditions hold:

(1) P_1 is true;

(2) For every positive integer k, if P_k is true, then P_{k+1} is true.

Condition (2) is often misunderstood; people think that it assumes what they are trying to prove. When you use mathematical induction, you are trying to prove: $(\forall n)(P_n$ is true). Condition (2) says, "$(\forall k)(P_k$ is true $\rightarrow P_{k+1}$ is true)". When you assert condition (2), you are speaking of a conditional statement. Like all "if ..., then" statements this is false if P_k is true and P_{k+1} is false, and it is true in all other cases, including those in which P_k is false. To prove Condition (2), you would assume P_k is true, for an arbitrary k, and prove that P_{k+1} is true, thus showing that the false case does not occur. It is important, of course, that you do not assume anything about k, other than it is a positive integer. A common error is to take a particular value, say 2, and prove that if P_2 is true, then P_3 is true.

The logic behind the Principle of Mathematical Induction is not difficult. Suppose conditions (1) and (2) hold, but that P_n is not always true. Let P_m be the first false statement, that is, m is the smallest positive integer for which the corresponding statement is false. Because of condition (1), m cannot be 1, so there are values of n less than m for which P_n is true. In particular, P_{m-1} is true because P_m was the first false statement. Then by condition (2), P_m must be true, a contradiction. It follows that all the statements in the sequence must be true.

EXERCISE. Use mathematical induction to prove each of the following:

1.* A fundamental property of logarithms says that if x and y are positive real numbers, then $\log(xy) = \log(x) + \log(y)$. (You can take these to be logarithms to any base you would like.) Prove that for every positive integer n, $\log(x^n) = n \log(x)$.

2. Prove that for every positive integer n, $2^n > n$.

3. Let x be a fixed positive real number. Prove that for every positive integer n, $(1 + x)^n \geq 1 + nx$.

Appendix S: Solutions

In this appendix I have written out proofs of selected theorems and exercises (those marked in the book with an asterisk). These proofs can be used in two ways.

You can compare your proofs with those given here to see if your work is correct. This is not, however, as simple as it sounds because in many cases there is more than one correct proof. If your proof is not the same as mine, do not assume that you are wrong but, at least, view your proof with suspicion. Go over it carefully to check for improper assumptions and logical errors. Look to see if you actually used all the given statements. If you do not find any errors, you probably are right, but you should show your proof to someone else, preferably your teacher. This careful scrutiny of your work is useful and important and you should do it with all your proofs. Remember, in your later work you will have to be the judge of your own work.

The second way this appendix can be used is to speed up the pace of the work. You can carefully read and study my proofs without trying to construct your own proofs and restrict your efforts to those theorems and exercises not included in the appendix. If you do this, you will be able to cover the later chapters of this book and so get a better foundation in the theory of metric spaces. If you are already competent in constructing proofs or as you develop such competence, this would be the better way to use the appendix.

In this appendix at first I write out proofs in detail, numbering the steps and giving reasons. Later I omit the more obvious reasons and eventually I abandon the line numbering. Most proofs in mathematics books and papers are not written in a step-wise fashion, but in a narrative style with trivial steps omitted or compressed. You have to become familiar with this style and eventually use it yourself. Therefore, I have tried in this appendix to lead you to this way of reading and writing proofs. You should remember, however, that when you are confused in a proof, it is a good idea to try writing it in a step-by-step fashion. Also you should remember to have clearly in mind what is "given" and what is "to be proved." Many times you will not find these explicitly stated in proofs that you will be reading, but you should always supply them.

Chapter I.   Section 2.

1.   Prove: $(A \times B) \cap (C \times D) = (A \cap C) \times (B \cap D)$.

    (i) Given: $(x,y) \in (A \times B) \cap (C \times D)$

        Prove: $(x,y) \in (A \cap C) \times (B \cap D)$

        Proof: 1. $(x,y) \in (A \times B) \cap (C \times D)$ (Given)

                2. $(x,y) \in A \times B$ and $(x,y) \in C \times D$
                                          (Def. of intersection)

                3. $(x \in A$ and $y \in B)$ and $(x \in C$ and $y \in D)$
                                          (Def. of product)

                4. $(x \in A$ and $x \in C)$ and $(y \in B$ and $y \in D)$
                                          (Rearrangement)

                5. $(x \in A \cap C)$ and $(y \in B \cap D)$
                                            (Def. of intersection)

                6. $(x,y) \in (A \cap C) \times (B \cap D)$
                                            (Def. of product)

Comment. The meaning of the phrase "rearrangement" should be obvious. In a more formal sense it is based on properties of the connective "and". You could use truth tables to prove that "and" is commutative, that is, (P and Q) is logically equivalent to (Q and P), and associative, that is, ((P and Q) and R) is logically equivalent to (P and (Q and R)). I will usually assume that such relatively trivial ideas are obvious and will not interrupt the proofs to explain them.

    (ii) Given: $(x,y) \in (A \cap C) \times (B \cap D)$

         Prove: $(x,y) \in (A \times B) \cap (C \times D)$

         Proof: Just reverse the steps of the preceding proof. []

4. Prove: $c(\bigcup_{\gamma \in \Gamma} A_\gamma) = \bigcap_{\gamma \in \Gamma} (cA_\gamma)$ and $c(\bigcap_{\gamma \in \Gamma} A_\gamma) = \bigcup_{\gamma \in \Gamma} (cA_\gamma)$.

I will prove only the first equation. The second equation can be proved in a similar fashion, or you can use the trick I mentioned in the proof of Exercise 1.3 in Chapter 0.

    (i) Given: $x \in c(\bigcup_{\gamma \in \Gamma} A_\gamma)$

        Prove: $x \in \bigcap_{\gamma \in \Gamma} (cA_\gamma)$

        Proof: 1. $x \in c(\bigcup_{\gamma \in \Gamma} A_\gamma)$ (Given)

                   2. not-$(x \in \bigcup_{\gamma \in \Gamma} A_\gamma)$ (Def. of complement)

                   3. not-$((\exists \gamma)(x \in A_\gamma))$ (Def. of union)

                   4. $(\forall \gamma)$not-$(x \in A_\gamma)$ (Rule of negation)

                   5. $(\forall \gamma)(x \in cA_\gamma)$ (Def. of complement)

                   6. $x \in \bigcap_{\gamma \in \Gamma} (cA_\gamma)$ (Def. of intersection)

    (ii) Given: $x \in \bigcap_{\gamma \in \Gamma} (cA_\gamma)$

         Prove: $x \in c(\bigcup_{\gamma \in \Gamma} A_\gamma)$

         Proof: Just reverse the steps of the preceding proof. []

Chapter I. Section 3.

<u>Theorem</u> 3.1 (i). $f(\emptyset) = \emptyset$

Use a proof by contradiction, that is, assume $f(\emptyset)$ is not empty and derive a contradiction.

  Proof: 1. $f(\emptyset) \neq \emptyset$ (Assumption)

     2. $(\exists y)(y \in f(\emptyset))$ (Def. of empty set)

     3. $(\exists y)(\exists x)(x \in \emptyset \text{ and } y = f(x))$

               (Def. of image)

     4. Contradiction since $x \in \emptyset$ is false.

<u>Theorem</u> 3.2 (vi). $f^{\leftarrow}(cB) = c(f^{\leftarrow}(B))$

 (i) Given: $x \in f^{\leftarrow}(cB)$

   Prove: $x \in c(f^{\leftarrow}(B))$

   Proof: 1. $x \in f^{\leftarrow}(cB)$ (Given)

      2. $f(x) \in cB$ (Def. of inverse image)

      3. $f(x) \notin B$ (Def. of complement)

      4. $x \notin f^{\leftarrow}(B)$ (Def. of inverse image)

      5. $x \in c(f^{\leftarrow}(B))$ (Def. of complement)

There is now a second half to the proof, but this is done by reversing the sequence of steps in the preceding proof. []

<u>Theorem</u> 3.3 (i). $f(f^{\leftarrow}(B)) \subseteq B$

 Given: $y \in f(f^{\leftarrow}(B))$

 Prove: $y \in B$

 Proof: 1. $y \in f(f^{\leftarrow}(B))$ (Given)

    2. $(\exists x)(x \in f^{\leftarrow}(B) \text{ and } y = f(x))$

              (Def. of image)

    3. $(\exists x)(f(x) \in B \text{ and } y = f(x))$

              (Def. of inverse image)

    4. $y \in B$ []

Comment. Line 4 follows from line 3 by substitution. Since $y = f(x)$, you can substitute y for $f(x)$ in $f(x) \varepsilon B$. The quantifier $(\exists x)$ can be dropped because there is no explicit occurrence of x in the result.

Chapter I. Section 4.

Theorem 4.1 (i). $f(f^{\leftarrow}(B)) = B$ for every subset B of Y iff f is surjective.

By Theorem 3.3(i), $f(f^{\leftarrow}(B)) \subseteq B$, so you must prove the reverse inclusion under the assumption that f is surjective.

    Given: f is surjective

              $B \subseteq Y$

              $y \varepsilon B$

Prove: $y \varepsilon f(f^{\leftarrow}(B))$

Proof: 1. $y \varepsilon B$

        2. $(\exists x)(x \varepsilon X$ and $y = f(x))$ (f is surjective)

        3. $(\exists x)(x \varepsilon f^{\leftarrow}(B)$ and $y = f(x))$

                                      (Def. of inverse image)

        4. $y \varepsilon f(f^{\leftarrow}(B))$ (Def. of image)

To prove the converse, assume that the condition of the theorem holds and take, in particular, $B = Y$.

    Given: $f(f^{\leftarrow}(Y)) = Y$

    Prove: f is surjective

    Proof: 1. $f^{\leftarrow}(Y) = X$ (Theorem 3.2 (ii))

            2. $f(f^{\leftarrow}(Y)) = Y$

            3. $f(X) = Y$ (Substitution)

            4. f is surjective []   (Def. of surjective)

Theorem 4.2. If $f: X \longrightarrow Y$, $g: Y \longrightarrow Z$ and $h: Y \longrightarrow Z$ are mappings such that f is a surjection and $gf = hf$, then $g = h$.

Given: f is surjective
gf = hf
y ε Y

Prove: g(y) = h(y)

Proof: 1. y ε Y
2. (∃x)(x ε X and y = f(x))   (f is surjective)
3. (gf)(x) = (hf)(x)   (gf = hf)
4. g(f(x)) = h(f(x))
5. g(y) = h(y)   []   (Substitution)

Theorem 4.5 (ii). f(A ∩ A') = f(A) ∩ f(A') for all subsets A and A' of X iff f is injective.

By Theorem 3.1 (v), f(A ∩ A') ⊆ f(A) ∩ f(A'), so you must prove the reverse inclusion under the assumption that f is injective.

Given: f is injective
y ε f(A) ∩ f(A')

Prove: y ε f(A ∩ A')

Proof: 1. y ε f(A) ∩ f(A')
2. y ε f(A) and y ε f(A')
3. (∃x)(x ε A and y = f(x)) and (∃x')(x' ε A' and y = f(x'))
4. f(x) = f(x')
5. x = x'   (f is injective)
6. (∃x)(x ε A and x ε A' and y = f(x))
7. y ε f(A ∩ A')

To prove the converse, take the special case of the condition of the theorem in which the subsets are singletons.

Given: f(x) = f(x')
f({x} ∩ {x'}) = f({x}) ∩ f({x'})

Prove: x = x'

Proof: 1. $f(x) \varepsilon f(\{x\})$ and $f(x') \varepsilon f(\{x'\})$
         2. $f(x) = f(x')$
         3. $f(\{x\}) \cap f(\{x'\}) \neq \emptyset$
         4. $f(\{x\} \cap \{x'\}) \neq \emptyset$
         5. $\{x\} \cap \{x'\} \neq \emptyset$ (Theorem 3.1 (i))
         6. $x = x'$ []

Theorem 4.7. If $gf = i_X$, then f is injective.

Given: $gf = i_X$
         $f(x) = f(x')$
Prove: $x = x'$
Proof: 1. $f(x) = f(x')$
         2. $g(f(x)) = g(f(x'))$ (Uniqueness of images)
         3. $(gf)(x) = (gf)(x')$
         4. $i_X(x) = i_X(x')$
         5. $x = x'$ []

Chapter I. Section 5.

Theorem 5.1. If $f: X \longrightarrow Y$ is a bijection, then $f^{-1}: Y \longrightarrow X$ is a mapping.

(i) First prove that Y is the domain for the rule f^{-1}.

Given: $y \varepsilon Y$
Prove: $(\exists x)(x \varepsilon X$ and $f^{-1}(y) = x)$
Proof: 1. $y \varepsilon Y$
         2. $(\exists x)(x \varepsilon X$ and $f(x) = y)$ (f is surjective)
         3. $(\exists x)(x \varepsilon X$ and $f^{-1}(y) = x)$ (Def. of f^{-1})

(ii) Next prove that X is the codomain of f^{-1}, but this is clear from the definition of the rule f^{-1}.

(iii) Lastly, you must prove that $f^{-1}(y)$ is a unique element of X. To do this, assume that $f^{-1}(y)$ has two values and prove that they are equal.

Given: $f^{-1}(y) = x_1$ and $f^{-1}(y) = x_2$

Prove: $x_1 = x_2$

Proof:
1. $f^{-1}(y) = x_1$ and $f^{-1}(y) = x_2$
2. $f(x_1) = y$ and $f(x_2) = y$ (Def. of f^{-1})
3. $f(x_1) = f(x_2)$
4. $x_1 = x_2$ []   (f is injective)

Comment. You might think that I could have concluded the proof of this last part right after line 1 by saying that x_1 and x_2 are equal because they are both equal to $f^{-1}(y)$. The error in such an argument is that you do not know if $f^{-1}(y)$ is a single element. On the other hand, line 3 follows from line 2 because $f(x_1)$ and $f(x_2)$ do both equal the single element y. This might become less mysterious if you take the example of $f: R \longrightarrow R$ defined by $f(x) = x^2$. If you try to apply the definition of f^{-1}, you would have to say that $f^{-1}(y)$ equals a square root of y. You could then have both 2 and -2 equal to square roots of 4, but they are not equal to each other.

Theorem 5.2. If $f: X \longrightarrow Y$ is a bijection, then $f^{-1}f = i_X$ and $ff^{-1} = i_Y$.

Clearly, $f^{-1}f$ is a mapping whose domain and codomain are both X, just as they are for i_X. To prove that $f^{-1}f = i_X$, you must just show that $(f^{-1}f)(x) = i_X(x)$ for any x in X.

Given: $x \in X$

Prove: $(f^{-1}f)(x) = i_X(x)$

Proof: Let $f(x) = y$ (Notation)

$f^{-1}(y) = x$ (Def. of f^{-1})

$(f^{-1}f)(x) = f^{-1}(f(x)) = f^{-1}(y) = x = i_X(x)$ []

Chapter II. Section 1.

EXERCISE 2. Prove: (R^2, d') is a metric space.

The proofs of (M1), (M2) and (M3) are straightforward, so I will not give them. To prove (M4) first prove the following lemma. (A lemma is just a theorem which is used primarily to prove another theorem.)

<u>Lemma.</u> For any real numbers a, b, c, and d
$$\max[a,b] + \max[c,d] \geq \max[a + c, b + d].$$

Proof:
1. $a \leq \max[a,b]$ and $c \leq \max[c,d]$ (Def. of max)
2. $a + c \leq \max[a,b] + \max[c,d]$
3. Similarly, $b + d \leq \max[a,b] + \max[c,d]$
4. $\max[a + c, b + d] \leq \max[a,b] + \max[c,d]$
                                          (Def. of max)

Now to prove (M4) let $p = (x_1, y_1)$, $q = (x_2, y_2)$, and $r = (x_3, y_3)$.

$d'(p,q) + d'(q,r) = \max[|x_1 - x_2|, |y_1 - y_2|] + \max[|x_2 - x_3|, |y_2 - y_3|]$

$\geq \max[|x_1 - x_2| + |x_2 - x_3|, |y_1 - y_2| + |y_2 - y_3|]$ (Lemma)

$\geq \max[|x_1 - x_2 + x_2 - x_3|, |y_1 - y_2 + y_2 - y_3|]$
                                               (Property of absolute values)

$$= \max[|x_1 - x_3|, |y_1 - y_3|]$$
$$= d'(p,r) \quad []$$

EXERCISE 4. Prove that (X,d) is a metric space, where X is the set of bounded real valued functions defined on $I = \{x \in R \mid 0 \leq x \leq 1\}$ and $d(f,g) = \sup \{|f(x) - g(x)| \mid x \in I\}$.

First you should prove that $\{|f(x) - g(x)| \mid x \in I\}$ has a supremum. To do this it is only necessary to prove that it has an upper bound. (See the appendix to this section.)

    Given: $f \in X$ and $g \in X$
    Prove: $\{|f(x) - g(x)| \mid x \in I\}$ has an upper bound
    Proof: $(\exists K)(|f(x)| \leq K$ for all x in $I)$ (f is bounded)
              $(\exists L)(|g(x)| \leq L$ for all x in $I)$ (g is bounded)
              $|f(x) - g(x)| \leq |f(x)| + |g(x)|$ for all x in I
              $|f(x) - g(x)| \leq K + L$ for all x in I
              $\{|f(x) - g(x)| \mid x \in I\}$ has an upper bound

Properties (M1) and (M3) are obviously true and $d(f,f) = 0$. To prove the remaining part of (M2), prove the contrapositive: if $f \neq g$, then $d(f,g) \neq 0$.

    Given: $f \neq g$
    Prove: $d(f,g) \neq 0$
    Proof: $(\exists x)(x \in I$ and $f(x) \neq g(x))$
              $(\exists x)(x \in I$ and $|f(x) - g(x)| > 0)$
              0 is not an upper bound of $\{|f(x) - g(x)| \mid x \in I\}$
              $\sup \{|f(x) - g(x)| \mid x \in I\} > 0$
              $d(f,g) \neq 0$

To prove (M4) take f, g, and h in X and let
        $A = \{|f(x) - g(x)| \mid x \in I\}$

$B = \{|g(x) - h(x)| \mid x \in I\}$

$C = \{|f(x) - h(x)| \mid x \in I\}$.

You have to prove that sup C $\leq$ sup A + sup B.

Proof: For any x in I,

$|f(x) - h(x)| = |(f(x) - g(x)) + (g(x) - h(x))|$

$\leq |f(x) - g(x)| + |g(x) - h(x)|$

$\leq$ sup A + sup B

sup A + sup B is an upper bound of $\{|f(x) - h(x)| \mid x \in I\}$

sup C $\leq$ sup A + sup B   []

Chapter II. Section 2.

Theorem 2.2. If p and q are distinct points of a metric space, then there exist two disjoint open balls, one with center at p and the other with center at q.

Let $d(p,q) = r > 0$. Take $B(p;r/2)$ and $B(q;r/2)$. You have to show that these open balls are disjoint, that is, have an empty intersection. Use a proof by contradiction.

Given: $x \in B(p;r/2) \cap B(q;r/2)$

Prove: a contradiction

Proof: $d(p,x) < r/2$ and $d(q,x) < r/2$

$d(p,q) \leq d(p,x) + d(x,q)$ (M4)

$= d(p,x) + d(q,x)$ (M3)

$< r/2 + r/2 = r$

$r < r$, a contradiction   []

Comment. You do not have to use the radius r/2 for the balls; any smaller radius would also work.

Chapter II. Section 3.

Theorem 3.2. An open ball is an open set.

Let $B(p;r)$ be the open ball. For each x in $B(p;r)$ you have to find an open ball with center at x that is contained in $B(p;r)$.

Given: $x \in B(p;r)$

Prove: $(\exists s)(B(x;s) \subseteq B(p;r))$

Proof: Let $s = r - d(p,x)$. Since $d(p,x) < r$, s is positive. To show that $B(x;s) \subseteq B(p;r)$, start with $y \in B(x;s)$ and prove that y is in $B(p;r)$.

$d(y,x) < s = r - d(p,x)$, so $d(y,x) + d(p,x) < r$. Then $d(y,p) \leq d(y,x) + d(x,p) < r$, so $y \in B(p;r)$. []

Theorem 3.4. A subset of a metric space is open iff it is a union of a family of open balls.

A union of open balls is open by Theorems 3.2 and 3.3, so it is only necessary to prove that an open set G is a union of a family of open balls.

For each p in G there is a real number r_p such that $B(p;r_p) \subseteq B$. Take the family $(B(p;r_p) \mid p \in G)$ and prove that the union of this family equals G.

Since $B(p;r_p) \subseteq G$ for each p in G, the union is clearly contained in G. Conversely, start with x in G and prove that x is in $\underset{p \in G}{\cup} B(p;r_p)$. But since $x \in G$, there exists r_x such that $B(x;r_x) \subseteq G$, so $x \in B(x;r_x) \subseteq \underset{p \in G}{\cup} B(p;r_p)$. This proves that $G = \underset{p \in G}{\cup} B(p;r_p)$. []

Comment. There is some ambiguity in the definition of the family, because for each p in G there are infinitely many real numbers that could serve as the radius of a ball with center at p that is contained in G. To avoid this, some rule should be given so that the actual real number that is used is specified. This can be done, but it requires some further study of sets of real numbers that I do not want to go into.

Chapter II. Section 4.

Theorem 4.1. A subset of a metric space is closed if its complement is open.

Proof: The proof consists of a sequence of "if and only if" statements.

F is closed $\leftrightarrow$ F contains all its accumulation points
$\leftrightarrow$ All points of cF are not accumulation points of F
$\leftrightarrow$ $(\forall p)(p \in cF$ and p is not an accumulation point of F$)$
$\leftrightarrow$ $(\forall p)(\exists r)(p \in cF$ and $(B(p;r) - \{p\}) \cap F = \emptyset)$
$\leftrightarrow$ $(\forall p)(\exists r)(p \in cF$ and $B(p;r) \cap F = \emptyset)$
$\leftrightarrow$ $(\forall p)(\exists r)(p \in cF$ and $B(p;r) \subseteq F)$
$\leftrightarrow$ cF is open   []

Theorem 4.5. If p is an accumulation point of a subset A of a metric space X, then for any positive real number r, $B(p;r)$ contains infinitely many distinct points of A.

Proof: Prove the contrapositive. Suppose $B(p;r) \cap A$ is finite, say it equals $\{a_1, a_2, \ldots, a_n\}$. If one of these numbers happens to be p, just discard it. Let s be the smallest of the numbers $d(p,a_i)$, $i = 1, \ldots, n$. Then $(B(p;s) - \{p\}) \cap A = \emptyset$, which shows that p is not an accumulation point of A. []

Comment. The finiteness of $B(p;r) \cap A$ is used in defining s. If this set were infinite, the set of all the distances $d(p,a_i)$ might not have a minimum.

Chapter II. Section 5.

Theorem 5.2. A point p is an element of $\bar{A}$ iff for every positive real number r, $B(p;r) \cap A \neq \emptyset$.

Proof: Suppose $p \in \bar{A}$. If $p \in A$, obviously $B(p;r) \cap A \neq \emptyset$, regardless of the value of r, since p is in this intersection. Therefore, assume that p is an accumulation point of A. For any positive r, $(B(p;r) - \{p\}) \cap A \neq \emptyset$. However, $B(p;r) \cap A \supseteq (B(p;r) - \{p\}) \cap A$, so $B(p;r) \cap A \neq \emptyset$.

For the converse I will prove the contrapositive: if $p \notin \bar{A}$, then $(\exists r)(B(p;r) \cap A = \emptyset)$. Since $p \notin \bar{A}$, it follows that $p \notin A$ and p is not an accumulation point of A. Therefore, $(\exists r)((B(p;r) - \{p\}) \cap A = \emptyset)$. Since $p \notin A$, it then is true that $B(p;r) \cap A = \emptyset$ for this value of r. []

Theorem 5.3 (ii). $\bar{A}$ is closed.

Proof: I will show that $Cl_X(\bar{A}) \subseteq \bar{A}$, which by the comment after Theorem 5.1 will prove that $\bar{A}$ is closed. Take $p \in Cl_X(\bar{A})$. Then for every r, $B(p;r) \cap \bar{A} \neq \emptyset$. Let $q \in B(p;r) \cap \bar{A}$ and take $s = r - d(p,q)$, which is a positive real number. Since $q \in \bar{A}$, $B(q;s) \cap A \neq \emptyset$, so take $x \in B(q;s) \cap A$. Then
$$d(p,x) \leq d(p,q) + d(q,x) < d(p,q) + s = r,$$
so $x \in B(p;r) \cap A$. This shows that $B(p;r) \cap A \neq \emptyset$, where r is any positive real number. Therefore, $p \in \bar{A}$, so $Cl_X(\bar{A}) \subseteq \bar{A}$. []

Theorem 5.3 (iv). $\bar{A}$ is the intersection of all the closed subsets of X which contain A.

Proof: Let $C = \cap\{F \mid F$ is closed and $F \supseteq A\}$. Then C is closed by Theorem 4.3 (i) and $A \subseteq C$. By Theorem 5.3 (iii) it follows that $\bar{A} \subseteq C$. On the other hand, $\bar{A}$ is itself a closed set which contains A, so it is in the family $\{F|\ F$ is closed and $F \supseteq A\}$. Therefore, it contains the intersection of this family, so $\overline{A} \supseteq C$. The two inclusions prove that $\overline{A} = C$. []

Chapter II. Section 6.

Theorem 6.4. A nonempty subset of a metric space X is bounded iff there is a point p in X such that A is contained in some closed ball with center at p.

Proof: Suppose A is bounded. If A is a singleton, the conclusion of the theorem is clearly true. Therefore, assume A is not a singleton, so $\delta(A) > 0$. Let q be some definite point of A and p be any point of X. Take $r = d(p,q) + \delta(A)$; then $r > 0$. I claim that $A \subseteq B^*(p;r)$. Take any x in A. Then $d(p,x) \leq d(p,q) + d(q,x) \leq d(p,q) + \delta(A) = r$, so $x \in B^*(p;r)$, as required.

Conversely, suppose $A \subseteq B^*(p;r)$. $B^*(p;r)$ is bounded by Theorem 6.1 (iii), so A is bounded by Theorem 6.1 (i). []

Chapter II. Section 7.

Theorem 7.2. Let Y be a subspace of the metric space X and $A \subseteq Y$. Then A is open in Y iff there exists a set G which is open in X such that $A = G \cap Y$.

Proof: A is open in Y $\leftrightarrow$ A is the union of open balls in Y

$\leftrightarrow A = \bigcup_{p \in A} B_Y(p;r_p)$ (See proof of Theorem 3.4)

$\leftrightarrow A = \bigcup_{p \in A} (B_X(p;r_p) \cap Y)$ (Theorem 7.1)

$\leftrightarrow A = (\bigcup_{p \in A} B_X(p;r_p)) \cap Y$ (Exercise I 2.5)

$\leftrightarrow A = G \cap Y$, where $G = \bigcup_{p \in A} B_X(p;r_p)$ is open in X. []

Theorem 7.4. Let Y be a subspace of the metric space X and $A \subseteq Y$. Then A is closed in Y iff there exists a set F which is closed in X such that $A = F \cap Y$.

Proof: First observe the following result about complements, which is easily proved: if $A \subseteq Y \subseteq X$, then $c_Y(A) = c_X(A) \cap Y$.

$\quad$ A is closed in Y $\leftrightarrow c_Y(A)$ is open in Y

$\quad\quad \leftrightarrow (\exists G)(G$ is open in X and $c_Y(A) = G \cap Y)$

$\quad\quad \leftrightarrow (\exists G)(G$ is open in X and $A = c_Y(G \cap Y))$.

However, $c_Y(G \cap Y) = c_X(G \cap Y) \cap Y \quad$ (See result above.)

$\quad\quad\quad\quad\quad = (c_X(G) \cup c_X(Y)) \cap Y \quad$ (Exercise I 2.4)

$\quad\quad\quad\quad\quad = [c_X(G) \cap Y] \cup [c_X(Y) \cap Y] \quad$ (Exercise I 2.5)

$\quad\quad\quad\quad\quad = [c_X(G) \cap Y] \cup \emptyset = c_X(G) \cap Y$.

So, A is closed in Y $\leftrightarrow (\exists G)(G$ is open in X and $A = c_X(G) \cap Y)$

$\quad\quad \leftrightarrow (\exists F)(F$ is closed in X and $A = F \cap Y)$, where $F = c_X(G)$. []

Chapter II. Section 8.

Theorem 8.1. If A is a subset of the metric space X, then
(i) $\overline{A} = c(Int_X(cA))$ and (ii) $A° = c(\overline{cA})$.

Proof of (i): $p \in c\overline{A} \leftrightarrow p \notin \overline{A}$

$\quad\quad\quad\quad \leftrightarrow (\exists r)(B(p;r) \cap A = \emptyset)$

$\quad\quad\quad\quad \leftrightarrow (\exists r)(B(p;r) \subseteq cA)$

$\quad\quad\quad\quad \leftrightarrow p \in (cA)°$.

This proves that $c\overline{A} = Int_X(cA)$. Take complements of both sides of this equation to get (i).

Proof of (ii): $Cl_X(cA) = c[Int_X(c(cA))] \quad$ (by (i))

$\quad\quad\quad\quad\quad\quad = c(Int_X(A))$.

Now take complements of both sides of this equation to get (ii). []

Chapter II. Section 9.

<u>Theorem</u> 9.3. (i) A is open iff A ∩ Bd(A) = ∅.

Proof: Suppose A is open and p is any point in A. (If A is empty, the result is obviously true.) Then there is an r such that B(p;r) ⊆ A, that is, B(p;r) ∩ cA = ∅. Therefore, p is not in Bd(A), so A ∩ Bd(A) = ∅.

Conversely, suppose A is not open. Then there is a point p which is in A - A°, so for any r, B(p;r) ⊄ A, that is, B(p;r) ∩ cA ≠ ∅. On the other hand, p ε B(p;r) ∩ A, so B(p;r) ∩ A ≠ ∅. Hence, p ε Bd(A) and so A ∩ Bd(A) ≠ ∅. []

<u>Theorem</u> 9.4. (ii) A° = A - Bd(A).

Proof: If p ε A°, then p ε A and (∃r)(B(p;r) ⊆ A), that is, B(p;r) ∩ cA = ∅. Therefore, p ∉ Bd(A), so A° ⊆ A - Bd(A). For the opposite inclusion take p ε A - Bd(A), that is, p ε A and p ∉ Bd(A). For every r, p ε B(p;r), so B(p;r) ∩ A ≠ ∅. Since p ∉ Bd(A), there must exist an r such that B(p;r) ∩ cA = ∅. For this r, B(p;r) ⊆ A, so p ε A°. This proves A° ⊇ A - Bd(A). The two inclusions prove that A° = A - Bd(A). []

Chapter II. Section 10.

<u>Theorem</u> 10.1. A subset of the metric space X is dense in X iff the only closed set containing A is X.

Proof: Suppose A is dense and F is a closed set containing A. Then $\bar{A}$ ⊆ F. But $\bar{A}$ = X, so X ⊆ F, which gives X = F. Conversely, suppose the condition of the theorem holds. Take any p in X and prove that p ε $\bar{A}$. Suppose that p ∉ $\bar{A}$ and derive a contradiction. Since p ∉ $\bar{A}$, (∃r)(B(p;r) ∩ A = ∅). Therefore, for this value of r, A ⊆ cB(p;r). But this last set is closed, so cB(p;r) = X by the condition of the theorem. Take complements of both sides of this equation to get $B(p;r) = \emptyset$, which is a contradiction. []

Chapter III. Section 1.

Theorem 1.1. The mapping $f: X \longrightarrow Y$ is continuous at x_o iff for every open ball $B_Y(f(x_o);\varepsilon)$ in Y there exists an open ball $B_X(x_o;\delta)$ in X such that $f(B_X(x_o;\delta)) \subseteq B_Y(f(x_o);\varepsilon)$.

Proof: f is continuous at x_o

$\leftrightarrow$ $(\forall \varepsilon)(\exists \delta)(d(x,x_o) < \delta \rightarrow d'(f(x),f(x_o)) < \varepsilon)$

$\leftrightarrow$ $(\forall \varepsilon)(\exists \delta)(x \in B_X(x_o;\delta) \rightarrow f(x) \in B_Y(f(x_o);\varepsilon))$

$\leftrightarrow$ $(\forall \varepsilon)(\exists \delta)(f(B_X(x_o;\delta)) \subseteq B_Y(f(x_o);\varepsilon))$

This last line is equivalent to the condition of the theorem. []

Theorem 1.2. Let f be a mapping from the metric space X to the metric space Y. Then f is continuous iff for every open set G in Y, $f^{\leftarrow}(G)$ is an open set in X.

Proof: Suppose f is continuous. Take any open set G in Y and prove that $f^{\leftarrow}(G)$ is open. Take any x_o in $f^{\leftarrow}(G)$ and find an open ball with center at x_o that is contained in $f^{\leftarrow}(G)$. But $f(x_o)$ is in the open set G, so $(\exists \varepsilon)(B_Y(f(x_o);\varepsilon) \subseteq G)$. By the continuity of f there exists a positive number δ such that

$$f(B_X(x_o;\delta)) \subseteq B_Y(f(x_o);\varepsilon) \subseteq G .$$

This means that $B_X(x_o;\delta) \subseteq f^{\leftarrow}(G)$, so $f^{\leftarrow}(G)$ is open.

Conversely, suppose that the inverse image of any open set in Y is open in X. Take any x_o in X and any $\varepsilon > 0$. Since $B_Y(f(x_o);\varepsilon)$ is open in Y, $f^{\leftarrow}(B_Y(f(x_o);\varepsilon))$ is open in X. But $x_o \in f^{\leftarrow}(B_Y(f(x_o);\varepsilon))$, so $(\exists \delta)[B_X(x_o;\delta) \subseteq f^{\leftarrow}(B_Y(f(x_o);\varepsilon))]$. Therefore,

$f(B_X(x_0;\delta)) \subseteq B_Y(f(x_0);\varepsilon)$, so f is continuous at x_0. Since x_0 is an arbitrary point of X, this proves that f is continuous. []

Chapter III. Section 2.

Theorem 2.4. Let X and Y be metric spaces, A be a nonempty subset of X, and $f: X \longrightarrow Y$ be a mapping. If $f|A$ is continuous, then f is continuous at every interior point of A.

Proof: Let ε be a positive number and $x_0 \in A^\circ$. Then $(\exists r)(B_X(x_0;r) \subseteq A)$. Since $f|A$ is continuous, $(\exists \delta)[f(B_A(x_0;\delta)) = f(B_X(x_0;\delta) \cap A) \subseteq B_Y(f(x_0);\varepsilon)]$. Take $\delta' = \min[\delta, r]$. Then $B_X(x_0;\delta') \subseteq B_X(x_0;\delta) \cap B_X(x_0;r) \subseteq B_X(x_0;\delta) \cap A$, so $f(B_X(x_0;\delta')) \subseteq f(B_X(x_0;\delta) \cap A) \subseteq B_Y(f(x_0);\varepsilon)$. This proves that f is continuous at x_0. []

Theorem 2.5. Let f and g be continuous mappings from the metric space X to the metric space Y. If A is a dense subset of X and if $f|A = g|A$, then $f = g$.

Proof: Suppose there is an x_0 in $X - A$ such that $f(x_0) \neq g(x_0)$. By Theorem II 2.2, there exist ε_1 and ε_2 such that $B_Y(f(x_0);\varepsilon_1) \cap B_Y(g(x_0);\varepsilon_2) = \emptyset$. Since f and g are continuous, there exist δ_1 and δ_2 such that $f(B_X(x_0;\delta_1)) \subseteq B_Y(f(x_0);\varepsilon_1)$ and $g(B_X(x_0;\delta_2)) \subseteq B_Y(g(x_0);\varepsilon_2)$. Take $\delta = \min[\delta_1, \delta_2]$. Then $f(B_X(x_0;\delta)) \subseteq B_Y(f(x_0);\varepsilon_1)$ and $g(B_X(x_0;\delta)) \subseteq B_Y(g(x_0);\varepsilon_2)$. Since A is dense, $B_X(x_0;\delta) \cap A \neq \emptyset$ by Theorem II 10.2. Let a be a point in this intersection. Since $a \in A$, $f(a) = g(a)$, so $f(a)$ is in both $B_Y(f(x_0);\varepsilon_1)$ and $B_Y(g(x_0);\varepsilon_2)$, which is a contradiction because these sets have no points in common. []

Chapter III. Section 3.

EXERCISE 1. Let X be the interval $(0,1]$ with the subspace metric of R. Define the mapping $f: X \longrightarrow R$ by $f(x) = 1/x$. Prove that f is continuous at each point of X, but f is not uniformly continuous on X.

Proof: Let ε be a positive real number and x_o be a point of $(0,1]$. Take $\delta = \varepsilon(x_o)^2/(1 + \varepsilon x_o)$. Then if $|x - x_o| < \delta$, with $x \in X$,

$$\left|\frac{1}{x} - \frac{1}{x_o}\right| = \frac{|x_o - x|}{xx_o} < \frac{\delta}{xx_o} = \frac{\varepsilon(x_o)^2}{(1 + \varepsilon x_o)xx_o} = \frac{\varepsilon x_o}{(1 + \varepsilon x_o)x}$$

But $x > x_o - \delta = x_o - \dfrac{\varepsilon(x_o)^2}{(1 + \varepsilon x_o)} = \dfrac{x_o}{(1 + \varepsilon x_o)}$, so $\dfrac{1}{x} < \dfrac{(1 + \varepsilon x_o)}{x_o}$.

It follows that $\left|\dfrac{1}{x} - \dfrac{1}{x_o}\right| < \varepsilon$, so f is continuous at x_o.

Comment. To see how I got the particular value of δ you might try working backwards.

To prove that f is not uniformly continuous, I will use a proof by contradiction. Suppose f is uniformly continuous and that δ corresponds to an arbitrary ε as in the definition. Take x_o to be a positive real number less than $\min[2\delta, \varepsilon^{-1}, 1]$. Then x_o is in X and

$$\left|x_o - \frac{x_o}{2}\right| = \left|\frac{x_o}{2}\right| < \frac{2\delta}{2} = \delta. \quad \text{However, for } x = x_o/2,$$

$$|f(x_o) - f(x)| = \left|\frac{1}{x_o} - \frac{2}{x_o}\right| = \left|-\frac{1}{x_o}\right| = \frac{1}{x_o} > \varepsilon$$

which is a contradiction. []

Chapter IV. Section 1.

EXERCISE 2. Prove that if (b_n) is a subsequence of (a_n) and (c_n) is a subsequence of (b_n), then (c_n) is a subsequence of (a_n).

Proof: There exist increasing mappings $\phi: \underline{N} \longrightarrow \underline{N}$ and $\psi: \underline{N} \longrightarrow \underline{N}$ such that $b_n = a_{\phi(n)}$ and $c_n = b_{\psi(n)}$ for every n. Then $c_n = b_{\psi(n)} = a_{\phi(\psi(n))} = a_{(\phi\psi)(n)}$. It is only necessary to prove that a composite $\phi\psi$ of two increasing mappings is an increasing mapping. Take i and j in $\underline{N}$ such that $i > j$. Then $(\phi\psi)(i) = \phi(\psi(i)) > \phi(\psi(j)) = (\phi\psi)(j)$, because $\psi(i) > \psi(j)$, since ψ is increasing, and ϕ is increasing. []

Lemma 1.1. Let $\phi: \underline{N} \longrightarrow \underline{N}$ be an increasing mapping. Then for every positive integer n, $\phi(n) \geq n$.

Proof: Use mathematical induction to prove that $\phi(n) \geq n$ for every positive integer n. (i) $\phi(1) \geq 1$ because $\phi(1) \in \underline{N}$ and 1 is the smallest element of $\underline{N}$. (ii) Assume $\phi(k) \geq k$ for some value of k. You have to prove that $\phi(k + 1) \geq k + 1$. But $k + 1 > k$ and ϕ is increasing, so $\phi(k + 1) > \phi(k) \geq k$. If an integer is greater than k, then it is certainly greater than or equal to $k + 1$. Therefore, $\phi(k + 1) \geq k + 1$, as required. []

EXERCISE 4. Let X be an infinite subset of the positive integers. Prove that there exists a sequence of distinct points of X. Take X as an infinite subset of any set and prove that there exists a sequence of distinct points of X.

Proof: Let $X \subseteq N$ be infinite. Define (a_n) by recursion. Take a_1 to be any element of X. Assume that $a_1, a_2, \ldots, a_n$ have been defined and are distinct points of X. Since X is infinite, $X - \{a_1, a_2, \ldots, a_n\} \neq \emptyset$. Choose a_{n+1} to be the smallest element of this set. Then a_{n+1} is different from a_1, a_2, ..., a_n. The recursion is then complete and this defines the sequence (a_n) of distinct points of X.

In the general case where X is an infinite subset of any set the recursion process proceeds in the same way, except you cannot choose a_{n+1} to be the smallest element of $X - \{a_1, a_2, ..., a_n\}$ because this set need not even have any ordering defined on it. However, since it is not empty, it contains elements and you merely choose one of them arbitrarily. Some mathematicians object to such an arbitrary choice in a recursion process, but most accept it as a legitimate process. This is part of an old controversy in set theory over what is called the "Axiom of Choice". If you are interested in philosophical problems and in questions about the foundations of mathematics, you can look up discussions about the "Axiom of Choice" in set theory texts. []

Chapter IV. Section 2.

Theorem 2.3. Let A be a subset of a metric space X. Then a point p of X is an accumulation point of A iff there exists a sequence (x_n) of points of A, none of which equals p, such that $\lim x_n = p$.

Proof: Assume p is an accumulation point of A. Define a sequence (x_n) by recursion as follows: Take x_1 to be any point of $(B(p;1) - \{p\}) \cap A$; this set is not empty because p is an accumulation point of A. Assume that the points x_1, x_2, ..., x_n have been chosen such that for $i = 1, 2, ..., n$, $x_i \in (B(p;1/i) - \{p\}) \cap A$. Since $(B(p;1/(n+1)) - \{p\}) \cap A$ is not empty, it is possible to choose x_{n+1} to be a point of this set. This defines the sequence (x_n) by recursion; all the points of the sequence are in A and none equals p. To show that $\lim x_n = p$, take any $\varepsilon > 0$ and let N be a positive integer such that $N > 1/\varepsilon$. Then if $n > N$, $1/n < 1/N < \varepsilon$, so

$x_n \in B(p;1/n) \subseteq B(p;\varepsilon)$, as required.

Conversely, suppose there is a sequence (x_n) of points of A, none of which equals p, such that $\lim x_n = p$. For any $\varepsilon > 0$ there is a positive integer N such that if $n > N$, $x_n \in B(p;\varepsilon)$. But then $(B(p;\varepsilon) - \{p\}) \cap A \neq \emptyset$, so p is an accumulation point of A. []

Chapter IV. Section 3.

Theorem 3.2. A point p of a metric space is a cluster point of the sequence (x_n) iff there is a subsequence $(x_{\phi(n)})$ of the sequence that converges to p.

Proof: Suppose p is a cluster point of (x_n). Construct an increasing mapping $\phi:\underline{N} \longrightarrow \underline{N}$ by recursion. Let $\phi(1)$ be the smallest subscript m such that $x_m \in B(p;1)$, that is, take the set of all the subscripts of points of the sequence that lie in $B(p;1)$ and choose the smallest one. Assume that $\phi(1), \phi(2), \ldots, \phi(n)$ have been defined in such a way that $\phi(1) < \phi(2) < \ldots < \phi(n)$ and $x_{\phi(i)} \in B(p;1/i)$, for $i = 1, 2, \ldots, n$. Define $\phi(n + 1)$ to be the smallest subscript m such that $m > \phi(n)$ and $x_m \in B(p;1/(n + 1))$. Since p is a cluster point of (x_n), there are infinitely many elements of the sequence in $B(p;1/(n + 1))$, so this choice is possible. This defines ϕ, which is obviously an increasing mapping. To prove that the subsequence $(x_{\phi(n)})$ converges to p, suppose $\varepsilon > 0$ is given and take N to be a positive integer such that $N > 1/\varepsilon$. Then if $n > N$, $x_{\phi(n)} \in B(p;1/n) \subseteq B(p;1/N) \subseteq B(p;\varepsilon)$, so $\lim x_{\phi(n)} = p$.

Conversely, suppose $(x_{\phi(n)})$ is a subsequence whose limit is p. To prove that p is a cluster point of (x_n), take any $\varepsilon > 0$ and any positive integer N. Since $\lim x_{\phi(n)} = p$ and $\varepsilon > 0$, there exists a positive integer M such that if $\phi(n) > M$, then $x_{\phi(n)} \in B(P;\varepsilon)$. Then there exists an integer m such that $\phi(m) > \max[M,N]$, for example $m = \max[M,N] + 1$ will work. (See Lemma 1.1.) But then $\phi(m) > N$ and $x_{\phi(m)} \in B(p;\varepsilon)$, which proves that p is a cluster point of (x_n). []

Chapter IV. Section 4.

Theorem 4.2. If (x_n) is a Cauchy sequence in a metric space X and if there is a cluster point p of the sequence in X, then (x_n) converges to p.

Proof: Let $\varepsilon > 0$ be given. Then there exists a positive integer N such that if $n > N$ and $m > N$, $d(x_m, x_n) < \varepsilon/2$. For $\varepsilon/2$ and N there is an $m > N$ such that $x_m \in B(p;\varepsilon/2)$, because p is a cluster point of the sequence. Fix this value of m. Then if $n > N$, $d(x_n, p) \leq d(x_n, x_m) + d(x_m, p) < \varepsilon/2 + \varepsilon/2 = \varepsilon$. This proves that $\lim x_n = p$. []

Theorem 4.5. The range of a Cauchy sequence is a bounded set.

Proof: Let (x_n) be a Cauchy sequence. Then there exists a positive integer N such that if $m > N$ and $n > N$, $d(x_m, x_n) < 1$. In particular, take $m = N + 1$, so that if $n > N$, $d(x_n, x_{N+1}) < 1$, that is, $x_n \in B(x_{N+1}; 1)$. Let $k = \max \{d(x_i, x_{N+1}) \mid 1 \leq i \leq N\}$. Then for all n, $x_n \in B(x_{N+1}; k + 1) \subseteq B^*(x_{N+1}; k + 1)$. Since the range of the sequence is contained in a closed ball, it is bounded. (Theorem II 6.4.) []

Chapter IV. Section 5.

Theorem 5.3. If every bounded infinite subset of a metric space X has an accumulation point in X, then X is complete.

Proof: Suppose every bounded infinite subset of X has an accumulation point in X. Let (x_n) be a Cauchy sequence in X. If the range of (x_n) is finite, then (x_n) is convergent by Corollary 4.4. If the range is infinite, it is bounded by Theorem 4.5, so by hypothesis it has an accumulation point. By Theorems 3.4 and 4.2, (x_n) converges. Since every Cauchy sequence in X converges, X is complete. []

Chapter V. Section 1.

Theorem 1.2. A metric space X is connected iff the only subsets of X which are both open and closed are ∅ and X.

Proof: Assume X is disconnected. Then there exist nonempty open sets A and B such that A ∩ B = ∅ and X = A ∪ B. Then A ≠ ∅ and A ≠ X. Clearly, A = cB, so A is closed, being the complement of an open set. Therefore, X contains a set which is both open and closed and is different from ∅ and X.

Conversely, assume that X has a subset A, different from ∅ and X, which is both open and closed. Let B = cA. Then A and B are nonempty open sets such that A ∩ B = ∅ and X = A ∪ B, so X is disconnected. []

Theorem 1.3. A nonempty subset Y of a metric space X is disconnected iff there exist open sets A and B in X with the following four properties: (i) A ∩ Y ≠ ∅, (ii) B ∩ Y ≠ ∅, (iii) (A ∩ B) ∩ Y = ∅, and (iv) Y ⊆ A ∪ B. Furthermore, the word "open" can be replaced by the word "closed".

Proof: Y is disconnected ↔ there exist open sets A_1 and B_1 in Y such that $A_1 \neq ∅$, $B_1 \neq ∅$, $A_1 \cap B_1 = ∅$ and $Y = A_1 \cup B_1$ ↔ there exist open sets A and B in X such that A ∩ Y ≠ ∅, B ∩ Y ≠ ∅, (A ∩ Y) ∩ (B ∩ Y) = ∅, and Y = (A ∩ Y) ∪ (B ∩ Y)

But $(A \cap Y) \cap (B \cap Y) = (A \cap B) \cap Y$. Also, $y = (A \cap Y) \cup (B \cap Y)$ iff $Y \subseteq A \cup B$. (Both of these results can be easily proved.) When these results are substituted into the above conditions, you get the statement of the theorem. If you apply Theorem 1.1, you get the conclusion about replacing "open" by "closed". []

<u>Theorem</u> 1.5. If $(A_\gamma \mid \gamma \in \Gamma)$ is a nonempty family of connected subsets of a metric space X and if $\cap_{\gamma \in \Gamma} A_\gamma \neq \emptyset$, then $\cup_{\gamma \in \Gamma} A_\gamma$ is connected.

Proof: Suppose $\cup_{\gamma \in \Gamma} A_\gamma \subseteq A \cup B$ and $(A \cap B) \cap (\cup_{\gamma \in \Gamma} A_\gamma) = \emptyset$, where A and B are open in X. Let $p \in \cap_{\gamma \in \Gamma} A_\gamma$. Then $p \in A$ or $p \in B$; say, $p \in A$. Then for each γ in Γ, $A_\gamma \cap A \neq \emptyset$. Since $A_\gamma \subseteq A \cup B$ and $A_\gamma \cap (A \cap B) = \emptyset$, the connectedness of A_γ gives that $A_\gamma \cap B = \emptyset$. This is true for each γ in Γ, so $(\cup_{\gamma \in \Gamma} A_\gamma) \cap B = \emptyset$. This proves that $\cup_{\gamma \in \Gamma} A_\gamma$ is connected.

<u>Chapter V. Section 3.</u>

<u>Theorem</u> 3.1. If $f: X \longrightarrow Y$ is a continuous mapping of metric spaces and if X is connected, then $f(X)$ is connected in Y.

Proof: Let A be a nonempty subset of $f(X)$ that is open and closed in $f(X)$. I will prove that $A = f(X)$ which shows that the only nonempty subset of $f(X)$ that is open and closed in $f(X)$ is $f(X)$ itself. It follows by Theorem 1.2 that $f(X)$ is connected.

There exist an open set G in Y and a closed set F in Y such that $A = G \cap f(X) = F \cap f(X)$. Then $f^{\leftarrow}(G)$ is open in X and $f^{\leftarrow}(F)$ is closed in X since f is continuous. I want to prove that $f^{\leftarrow}(G) = f^{\leftarrow}(F)$. This is done as follows:

$$x \in f^{\leftarrow}(G) \leftrightarrow f(x) \in G \leftrightarrow f(x) \in G \cap f(X) \leftrightarrow f(x) \in A$$
$$\leftrightarrow f(x) \in F \cap f(X) \leftrightarrow f(x) \in F \leftrightarrow x \in f^{\leftarrow}(F)$$

Then $f^{\leftarrow}(G)$ (which equals $f^{\leftarrow}(F)$) is both open and closed in the connected space X. This set is not empty because $A \neq \emptyset$, so it must be that $f^{\leftarrow}(G) = X$. Therefore, $f(X) = f(f^{\leftarrow}(G)) \subseteq G$ and so $A = G \cap f(X) = f(X)$, as required. []

Chapter VI. Section 1.

Theorem 1.4. A closed subset of a compact metric space is compact.

Proof: Let F be a closed subset of the compact metric space X and $(G_\gamma \mid \gamma \in \Gamma)$ be a family of open sets of X such that $F \subseteq \cup_{\gamma \in \Gamma} G_\gamma$. Take the family $(G_\gamma \mid \gamma \in \Gamma) \cup cF$. This is an open cover of X and since X is compact, there is a finite subfamily that covers X. If cF is in this subfamily, discard it. The union of the sets in the reamining finite family contains F, which proves that F is compact. []

Chapter VI. Section 3.

Theorem 3.3. If X is a compact metric space, then every infinite subset of X has an accumulation point in X.

Proof: Let A be an infinite subset of X which has no accumulation points in X. For each p in X there exists $r_p > 0$ such that $(B(p;r_p) - \{p\}) \cap A = \emptyset$, that is, $B(p;r_p) \cap A$ is either $\emptyset$ or $\{p\}$. $(B(p;r_p) \mid p \in X)$ is an open cover of X, so there exists a finite subset $\{p_1, p_2, \ldots, p_n\}$ of X such that $(B(p_i;r_i) \mid 1 \leq i \leq n)$ is still a cover. (I have written r_i in place of r_{p_i}.) Then
$$A = A \cap X = A \cap (\cup_{1 \leq i \leq n} B(p_i;r_i)) = \cup_{1 \leq i \leq n} (A \cap B(p_i;r_i)).$$
But each set in the union is either empty or a singleton, and since there are

finitely many sets, this shows that A is finite, a contradiction. []

APPENDIX M

1. Prove that for every positive integer n, $\log(x^n) = n \log(x)$.

Proof: P_n is the statement: $\log(x^n) = n \log(x)$.

(1) P_1 is obviously true.

(2) Assume P_k is true, that is, for some k, $\log(x^k) = k \log(x)$. Prove that P_{k+1} is true, that is, $\log(x^{k+1}) = (k + 1) \log(x)$. But, $\log(x^{k+1}) = \log(x \, x^k) = \log(x) + \log(x^k) = \log(x) + k \log(x) = (k + 1) \log(x)$, as required. []

Index

Accumulation point 41

Bijection 32

Bolzano-Weierstrass Theorem 83

Boundary 49

Boundary point 49

Bounded set 44

Cartesian product 16

Cauchy sequence 66

Closed ball 38

Closed set 41

Closure 43

Cluster point 64

Codomain 20

Compact set 76

Compact space 76

Complement 15

Complete space 67

Composite mapping 24

Connected space 69

Connected subset 70

Continuous mapping 53

Contrapositive 5

Convergent sequence 63

Counterexample 19

Cover 76

Dense 50

Diameter 44

Difference of sets 15

Disconnected space 69

Disconnected subset 70

Discrete metric 36

Divergent sequence 63

Domain 20

Embedding 32

Empty set 15

Epsilon net 80

Family 17

Finite intersection property 77

Heine-Borel Theorem 83

Identity mapping 30

Image 20

Indirect proof 5

Infinum 37

Injection 30

Interior 48

Interior point 48

Intermediate Value Theorem 73

Intersection of a family 17

Intersection of sets 15

Interval 71

Inverse image 20

Inverse mapping 32

Isometric spaces 57

Isometry 57

Limit of a sequence 63

Lower bound 37

Mapping 19

Metric 34

Metric space 35

Open ball 38

Open set 40

Precompact 80

Range 20

Restriction 20

Sequence 60

Sequentially compact 79

Sphere 38

Subcover 76

Subset 15

Subspace 46

Supremum 38

Surjection 29

Totally bounded 80

Unbounded set 44

Uniformly continuous 56

Union of a family 17

Union of sets 15

Upper bound 37

Usual metric on R 35

Usual metric on R^2 35

Universitext

Editors: F.W. Gehring, P.R. Halmos, C.C. Moore

Chern: Complex Manifolds Without Potential Theory
Chorin/Marsden: A Mathematical Introduction to Fluid Mechanics
Cohn: A Classical Invitation to Algebraic Numbers and Class Fields
Curtis: Matrix Groups
van Dalen: Logic and Structure
Devlin: Fundamentals of Contemporary Set Theory
Edwards: A Formal Background to Mathematics 1: Logic, Sets and Numbers
Edwards: A Formal Background to Mathematics 2: A Critical Approach to Elementary Analysis
Frauenthal: Mathematical Modeling in Epidemiology
Fuller: FORTRAN Programming: A Supplement for Calculus Courses
Gardiner: A First Course in Group Theory
Greub: Multilinear Algebra
Hájek/Havránek: Mechanizing Hypothesis Formation: Mathematical Foundations for a General Theory
Hermes: Introduction to Mathematical Logic
Kalbfleisch: Probability and Statistical Inference I/II
Kelly/Matthews: The Non-Euclidean, Hyperbolic Plane: Its Structure and Consistency
Kostrikin: Introduction to Algebra
Lu: Singularity Theory and an Introduction to Catastrophe Theory
Marcus: Number Fields
Meyer: Essential Mathematics for Applied Fields
Moise: Introductory Problem Courses in Analysis and Topology
Oden/Reddy: Variational Methods in Theoretical Mechanics
Reisel: Elementary Theory of Metric Spaces: A Course in Constructing Mathematical Proofs
Rickart: Natural Function Algebras
Schreiber: Differential Forms: A Heuristic Introduction